Georgia Wildmen: Exploring the History of Relict Hominoids in the Peach State

Erin Cain

Published by Erin Cain, 2024.

While every precaution has been taken in the preparation of this book, the publisher assumes no responsibility for errors or omissions, or for damages resulting from the use of the information contained herein.

GEORGIA WILDMEN: EXPLORING THE HISTORY OF RELICT HOMINOIDS IN THE PEACH STATE

First edition. February 8, 2024.

ISBN: 979-8224910373

Written by Erin Cain.

Table of Contents

Chapter One: Introduction...1

Chapter Two: Skeletons of Unusual Size.................................... 13

Chapter Three: Legends of the Cherokee.................................... 34

Chapter Four: Legends of the Okefenokee 60

Chapter Five: The Saga of Walker County 72

Chapter Six: Other Historical Accounts 94

Chapter Seven: Misidentifications and Hoaxes 145

Chapter Eight: Appendix.. 157

To my family.

For inspiring, supporting, and tolerating me.

Preface

Bigfoot has interested me for as long as I can remember, yet for the longest time, I never thought pursuing cryptozoology seriously was an option. In the age before content creators and influencers, I needed a respectable 9-to-5 career. I began a degree in wildlife biology until I felt a decision looming on the horizon between becoming a teacher, acquiring mountains of student debt, or moving to where it snows. That dismal prospect led to a decade-long detour in electrical engineering. It wasn't until I married and moved out of the city that I rekindled my interest in the strange and unexplained.

Once I began to study Bigfoot in earnest, I realized I had been wrong about my conception of the phenomenon. What I thought was one species across multiple continents was actually a plethora of relict hominoid species worldwide, including several in my home state of Georgia. This knowledge shook my understanding of biology, history, and mythology, and the contents of this book might do the same for you.

Chapter One: Introduction

Foundations

Bigfoot. Sasquatch. Yeti. Abominable Snowman. These are but a few of the thousands of terms used to describe human-ape hybrid beings across the globe. As a group, we can refer to these beings as relict hominoids: relict being a scientific term referring to a remnant species (or population) that persists despite all other related species (or populations) going extinct, and hominoid referring to members of the taxonomic superfamily Hominoidea, which includes humans, great apes, and lesser apes. Historically, humans and their ancestors have been called hominids, while great apes have been called pongids. In recent decades, this terminology has shifted as the great apes have been taxonomically reclassified and are now part of the family Hominidae alongside humans. The new verbiage refers to humans and their ancestors as hominins, while hominids encompass humans and great apes. This book uses the classic terminology of hominoids, hominids, and pongids. Primarily, it is a practical choice to use the outdated terms. By utilizing the vocabulary of our cryptozoological forefathers, we create consistency and avoid confusion between generations. Additionally, though the great apes have been taxonomically relocated closer to humans, the fact remains that great apes have 48 chromosomes while humans only have 46. This genetic binary must be respected, even if phenotypic overlap exists between the most ape-like humans and the most human-like apes.

Within the umbrella of relict hominoids, there are several varieties. For generations, researchers have been looking to differentiate and classify the distinct types of beings for which we have both physical and testimonial evidence. For example, within North America, the classic Sasquatch footprint of the Pacific Northwest looks much like a giant human footprint to the untrained eye—though it should be noted that Dr. Grover Krantz and Dr. Jeff Meldrum have shown conclusively that the skeletal anatomy displayed in these tracks is quite different from that of a modern human. On the other side of the Lower 48, in Florida, the footprints of the Skunk Ape clearly show a divergent hallux, better known as the big toe. This "hand-shaped" foot is a distinctly pongid characteristic found in gorillas and chimpanzees. Based on the footprint evidence alone, the Pacific Northwest Sasquatch and the Florida

Skunk Ape cannot be the same species; thus, there is a need for classification within the field of hominology, the name given by Russian researcher Dmitri Bayanov to the study of relict hominoids.

Various classification systems have been proposed over the last three-quarters of a century. In his book Abominable Snowmen, Ivan T. Sanderson divided ABSMs, as he called them, into four categories: Sub-Humans, Proto-Pigmies, Neo-Giants, and Sub-Hominids. Mark A. Hall divided relict hominoids into seven categories: Neo-Giants, True Giants, Yetis, Taller Hominids, Shorter Hominids, Little People, and Least Hominids. Building on Hall's work, Loren Coleman, in his book The Field Guide to Bigfoot, Yeti, and Other Mystery Primates Worldwide, divided relict hominoids into nine categories: Neo-Giants, True Giants, Marked Hominids, Neanderthaloids, Erectus Hominids, Proto-Pygmys, Unknown Pongids, Giant Monkeys, and Merbeings. Others have taken their own stab at dividing the types. Based on the testimonial evidence, Georgia has historically been home to four categories of relict hominoids: Eastern Bigfoots, Neanderthaloids, Proto-Pygmies, and Skunk Apes.

The Eastern Bigfoot type averages around 7 feet tall with short dark hair, usually brown, covering the whole body. They have large shoulder muscles, a neckless appearance, and a pointed head, which suggests the presence of a sagittal crest; thus, they have a jaw and skull structure similar to that of a gorilla. The same anatomical features can also be seen in Patty, the Pacific Northwest Sasquatch captured in the Patterson-Gimlin film. They are primarily nocturnal and have large eyes. Loren Coleman refers to the Eastern Bigfoot type as Marked Hominids, both because of the research of Mark A. Hall and because of their tendency for piebalding. This book uses Eastern Bigfoot because there is no historical evidence of piebalding in the Georgia population. The term Eastern Bigfoot also denotes a connection to the Pacific Northwest Sasquatch. Though some, like Coleman, would differentiate the Sasquatch of the Pacific Northwest from the Eastern Bigfoot as two distinct species, there is at least some relationship between the two types. The Eastern Bigfoot, as described in Georgia, is remarkably similar in appearance and behavior to the Pacific Northwest Sasquatch. North America is home to several species that show a wide range of physical appearance in size and coloration. For example, the common raccoon has 22 subspecies in North America and the Caribbean. These subspecies range from only 10 pounds to almost 60 pounds, from dark hair to blond hair, and from short-haired to long-haired. Until the taxonomic relationship can be determined based on genetic evidence, we cannot discount that the Pacific Northwest Sasquatch and the Eastern Bigfoot could be subspecies of the same specific lineage.

The Neanderthaloid type is the closest in appearance, and presumably genetics, to modern humans, Homo sapiens sapiens. On average, they are stronger, more muscular, and slightly taller than a typical human. They have long head hair and long facial hair with a thick coating of short body hair, and their hair tends to be more reddish brown than that of other relict hominoid types. They sometimes wear skins or ragged clothing acquired from humans and often use old tools and abandoned human lodgings. Though all primates have fingernails as opposed to the claws of canines and other animals, the fingernails of Neanderthaloids are often described as claws or talons by witnesses. Notably, Neanderthaloid types are assumed to be feral or savage humans when encountered historically. This assumption is not limited to accounts and legends from Georgia, either. The Bushmen Neanderthaloids of Northwestern Canada were also considered human by the First Nations. Athabaskan Bushman legends include sexual relations with human women and even the incorporation of some Bushmen into the local tribe. While anthropologists such as Ellen B. Basso declare the Bushmen to be human, as she considers the notion of relict hominoids outlandish, this conclusion ignores the body of legends that include non-human characteristics. For example, in The Ethnography of the Tanaina, anthropologist Cornelius Osgood documents the story of a man who lived with the Bushmen, or Bad Indians, for a year and then returned having "grown hair all over his body," which suggests the Bushmen had thick body hair just like the Neanderthaloids in Georgia. Of the relict hominoid types historically living in Georgia, the Neanderthaloid is the least likely to persist into the modern era precisely because they were presumed to be human. The ones that were discovered were shipped off to freak shows, prisons, and insane asylums, all places where maintaining a breeding population is impossible.

On the other end of the somewhat arbitrary spectrum between hominid and pongid is the Skunk Ape, the only type in Georgia we can safely categorize as pongid. The Skunk Ape is typically 5-6 feet tall with more ape-like proportions, having longer arms and shorter legs with a small pointed head and thick neck. Their hair is a consistent length all over but is described as shaggy, suggesting it is slightly longer than that of an Eastern Bigfoot. These nocturnal creatures tend towards the swamps and wooded bottomlands of the Southern United States, especially in Florida, where the Skunk Ape has become as legendary as the Pacific Northwest Sasquatch. Distinguishing characteristics of the Skunk Ape from other relict hominoid types are their divergent big toe and tendency to run quadrupedally.

The fourth type of relict hominoid witnessed in Georgia and the rarest in the testimonial evidence is the Proto-Pygmy type. Proto-Pygmies are generally petite, less than 5.5 feet tall, and slenderly built. Like the Neanderthaloid type, they have longer head hair than body hair, but their head hair is often described as mane-like or falling to the ground instead of only falling to the waist. They have tiny feet with pointed heels and are proficient at running, swimming, and climbing trees. Unfortunately, little can be gleaned about the Georgia population, as so few encounters were documented.

Ecology

The state of Georgia can be divided geographically into three general areas: North Georgia, Central Georgia, and South Georgia. North Georgia is the state's mountainous portion, comprising three geographical types. In the state's far northwest corner, Dade County sits on the edge of the Cumberland Plateau portion of the Appalachian Plateau. Historically, the geography isolated Dade County, and residents had to travel through Alabama or Tennessee to reach any other county in Georgia. The first road connecting Dade County to neighboring Walker County was not built until the 1930s. The rest of Northwest Georgia is part of the Ridge and Valley section of the Appalachian Highlands. The landscape has alternating high ridges and low valleys that run parallel from New York to Alabama. Northwest Georgia also boasts significant limestone and sandstone deposits, creating caves not present in other parts of the state. Northeast Georgia comprises the southern tip of the Blue Ridge section of the Appalachian Mountains. This area sees the most annual snowfall and rainfall in the state; it is home to the tallest point in Georgia, Brasstown Bald; and it serves as the southern gateway to the Appalachian Trail. Because of its terrain, Northeast Georgia is home to the most hydroelectric dams in the state.

Central Georgia is part of the Piedmont Province, with rolling foothills and red clay soil. This region has seen the most extreme population growth in the state in the last 150 years. For centuries, the Fall Line that divided the Piedmont of Central Georgia from the Coastal Plains of South Georgia was a natural limit to industrial expansion, as ships could not easily bypass the waterfalls and rapids that plague the rivers at that point. As a result, cities such as Columbus, Augusta, and Macon developed directly on the Fall Line. North and Central Georgia opened to industry only when technological development, particularly railroads and steam engines, allowed merchants to bypass such geographical impediments. The Atlanta Metro Area spans the border between North and Central Georgia. Today, it encompasses 28 of Georgia's 159 counties and over six of Georgia's ten million residents, but this was not always the case. Atlanta was a small rural town until its railroads became vital to the Confederacy's war effort and the subsequent occupying military government. Atlanta was selected as the new capital in 1868, and by 1880, it had surpassed Savannah as the largest city in Georgia. Most of the exponential population growth occurred in the second half of the 20th century, though, with the Metro Area tripling in population between 1950 and 1990. As a result of this population growth, the Atlanta Hartsfield-Jackson International Airport became the busiest airport in the world in 1998, a title it still holds. The rest of Central Georgia has also experienced considerable development. Columbus, Augusta, and Macon are Georgia's second, third, and fourth-largest cities.

South of the Fall Line lies the Coastal Plains of South Georgia. Southwest Georgia is part of the Gulf Coastal Plain, while Southeast Georgia is part of the Atlantic Coastal Plain. Unlike the crystalline geology of the Central Georgia Piedmont, where quartz, granite, and the like are predominant, South Georgia has sedimentary geologic structures that developed when the Atlantic Ocean covered South Georgia up to the Fall Line in the Mesozoic Period. Another unique feature of South Georgia is its coniferous forests. While North and Central Georgia have a temperate forest combination of shortleaf pine, loblolly pine, and hardwoods, particularly oaks, South Georgia forests are dominated by longleaf pine and slash pine with very few hardwoods.

Among the largest predators in the state are the American black bear, the bobcat, the coyote, and the gray and red foxes. Historically, there were also North American cougar and red wolf populations. Except for the Florida panther population, the cougar was extirpated from the eastern United States around the turn of the 20th century. The red wolf population declined in the middle of the 20th century due to habitat loss and hybridization with the coyote, whose range was expanding from the western United States.

The American black bear is the largest land mammal in Georgia and the most likely to be misidentified as a relict hominoid. Bears will fluctuate in weight considerably throughout the course of a year as they prepare for and then come out of hibernation, but they are typically about six feet in length and three feet tall at the shoulder. Their diet is that of a hypocarnivore. They primarily feed on plant material, such as berries, fruit, acorns, and grasses, but they will also consume insects, mammals, and whatever human-related food is opportunistically available. The current black bear population in the state is estimated to be around 5,000 individuals.

The Curation of the Stories Included

In 1912, a group of scientists from Cornell University, led by Dr. J. C. Bradley, went on a mission into the Okefenokee Swamp to document as many flora and fauna as possible. The following is a tale related by Dr. W. D. Funkhouser, a junior member of that expedition, which was included in the book *Georgia: Unfinished State* by Hal Steed, published in 1942.

"At the end of twenty days [Dr. Funkhouser and his team of four other scientists] came to a strip of dry land, of which there are several in the swamp. There they came upon a family of aborigines. In many ways, the doctor said, these people were like wild animals. There was an old woman, three sons, and two daughters. The children of these aborigines, he added, had intermarried and had eleven offspring of their own. All were characterized as degenerate weaklings, undernourished, full of hookworms. Tests showed that their blood was bad. A large graveyard near by was filled. The family's shelter was a crude lean-to built against a tree. The elders wore no clothing to speak of and the children were naked.

"Their speech, Dr. Funkhouser reported, was unintelligible, but he and his associates made a dictionary of it. Consider their amazement when they discovered that many of the words the family used had come down to them from Spenser, Chaucer and Shakespeare. This revelation is borne out by the quaint old forms and words still used by Georgia mountaineers in isolation.

"The aborigines had never heard of reading and writing and knew nothing of the world outside of their island. Was New York another island or a turpentine still? Only one of the men had ever been off the island. He had found the headwaters of the Suwanee River and had floated down it on a raft to a store where he swapped alligator skins and fur for food and other supplies. But they were religious persons and asked the scientists if they believed in a living God. Their senses were highly developed. They could smell rattlesnakes in the swamp and scent game like dogs. They saw things the scientists could not see, and they described animals and birds so accurately that the scientists could recognize them. They knew more about nature than he and his associates, Dr. Funkhouser had to admit."

While the story may seem extraordinary and might lead one's imagination to wander, this story is precisely why we must always remain alert and skeptical in reading historical accounts. One hundred to two hundred years is a long game of telephone, and journalists were as sensational in their writing then as journalists are today. Fiction was often presented as fact to sell newspapers, and fact-checking as we know it today was significantly more complicated, particularly in rural areas where families may only go to town once a month or so. Journalists and even learned scientists occasionally fell victim to merry pranksters and tall tales.

Alexander Stephens McQueen, who published *History of the Okefenokee Swamp* in 1926, believed this story resulted from merry pranksters. He felt Dr. Funkhouser fell victim to a "Georgia Cracker around the turpentine still" who was looking to pull one over on the "Yankee scientist" for entertainment. Given that Dr. Funkhouser claimed to have met and studied the family, he must have had at least some culpability in the yarn being spun.

Regardless, Dr. Funkhouser's story was false. In reality, the family living on Billy's Island was the family of Daniel Lee, who, with his wife, brought 16 children into this world and managed to raise 14 into adulthood, all of whom could read and write and none of whom participated in any inbreeding or intermarrying. They also wore clothes, lived in a proper cabin, and regularly traded with communities outside the swamp.

Corporate propaganda was the most likely motivation for the false representation of the family. The Daniel Lee family had been in lawsuits several times over the true ownership of Billy's Island. At the time of the Cornell expedition, the Hebard Cypress Company was again suing to extricate the island's inhabitants to make way for their timber harvesting operation.

Beyond simply sorting the fantastical from the factual, we also must be cautious when analyzing the language used in old accounts. Standard cultural references taken for granted then are now inside jokes lost to history and ripe for misinterpretation. Even seemingly straightforward terms such as "wild man" had various meanings. In addition to describing "wild men of the woods" who had the characteristics of relict hominoids, "wild man" was used to describe humans living in the woods due to mental illness, developmental disorders, or other issues. It was even used as a nickname for President Theodore Roosevelt, Georgia Governor Eugene Talmadge, and baseball player Ty Cobb.

Alternatively, we sometimes have to read between the lines in reports that seem mundane, particularly in the case of Neanderthaloid-type relict hominoids. It was a commonly held belief that man could devolve to a more savage state by simply living in the woods for too long, so reports of men covered in hair and unable to speak or understand human languages were classified as feral humans, not relict hominoids.

In curating the stories in this book, I have attempted to pay homage to legend while focusing on what I consider the most compelling reports. I have included stories that teeter on the edge of known versus unknown, but I have done my best to explain in the commentaries which reports are the most credible and insightful.

The sightings are limited to my home state of Georgia, except for the Cherokee legends. At the time when James Mooney was documenting the Cherokee mythology and culture, the Georgia population had already been removed to Oklahoma for several decades. However, Mooney conducted research with the Cherokee remnant in western North Carolina, now known as the Eastern Band of Cherokee Indians, whose Qualla Boundary territory is just 40 miles from the Georgia border. This information represents the best available on the mythology of the Cherokee within Georgia.

You will notice that some newspapers are referenced frequently, even for sightings in other areas of the state. For example, both the *Savannah Morning News* and the *Macon Telegraph* are commonly cited, partly because of the willingness of the writers to report on this type of phenomenon. Also,

Savannah and Macon are both major cities in Georgia, so we have better and more extensive records of the newspapers in these cities than we do in some of the smaller communities where the actual interactions between humans and hominoids took place. Some counties in Georgia have no archived newspapers from this period, so their stories must be told entirely from the perspective of other newspapers in the state.

As an additional note, it is essential to acknowledge that both spelling and manner of writing have evolved over the last two centuries. While I have not replaced any words—including language now deemed offensive—or altered the cadence of any of the accounts, I have made adjustments to spelling where obvious typos or outdated spellings would have made it more difficult for the reader to sort through the content of the reports. The intention is to preserve history—even the ugly bits—while making it accessible to the modern reader.

Chapter Two: Skeletons of Unusual Size

Standing on the Shoulders of Giants

While skeletons of giant proportions are not direct evidence of the existence of relict hominoids, they are certainly something we must reckon with. That said, giant skeleton finds of the 19th century are highly controversial. Many were discovered to be a hoax, many were falsely claimed to be a hoax, and even those submitted to scientific institutions were rarely examined in earnest. As Ivan T. Sanderson noted in his book *Abominable Snowmen*, many of the giant skeletons discovered across America disappeared "very often within the portals of some museum which had acknowledged receipt of the relic."

The following reports show that Georgia's long precolonial history includes more than just Native Americans of typical human stature. At the same time, we should not give too much importance to these reports. Without proper examination, it is impossible to determine if the bones are from a hoax, from a medical condition such as a pituitary tumor, or from something more mysterious.

Five reports of giant skeletons are included herein. One report is from Tybee Island, a barrier island off the coast of Savannah in Southeast Georgia. Two reports are from Indian mounds, one in Northwest Georgia and one in East Georgia. Both groups of mounds have been attributed to the South Appalachian Mississippian culture, which predated the Cherokees and Creeks in the area. One report is from a cave in West Georgia that seemed to have been used as a burial site before being covered by a landslide. The final report is from an island within the great Okefenokee Swamp.

The Rockmart Cave skeleton in West Georgia has the most potential of belonging to a relict hominoid. Unfortunately, only the skeleton at the Etowah Mounds received any formal scientific examination, and only minimally so.

Tybee Island

Location: Chatham County, Southeast Georgia

November 16, 1875 - Savannah Morning News (Savannah, Chatham County, GA)

We learn that the force under Capt. Brown, engaged in laying the tramway at Tybee Island, unearthed on Saturday last, at a point just east of McKenzie's house the skeleton of a giant. The skeleton measured eight feet in length. The discovery created considerable sensation among the "inhabitants" of the Island, and we are now prepared to receive some information on the subject from the knowing ones.

November 24, 1875 - Weekly Gwinnett Herald (Lawrenceville, Gwinnett County, GA)

A giant skeleton, eight feet in length, has been unearthed on Tybee Island.

Commentary:

Tybee Island is one of the barrier islands of Georgia. Overlooking the entrance to the Savannah River, Tybee served as the first line of defense for the port of Savannah in the Spanish-American War and both World Wars. Today, it is a popular tourist destination, both for travelers and Savannah residents, but 150 years ago, Tybee was practically a virgin landscape.

In 1875, Tybee Island was only just being settled. The Tybee Improvement Company began transforming the island's raw state into a resort destination in 1870. Still, the first lots weren't surveyed until 1873, and the first resort hotel didn't open its doors until 1876. The tramway, which carried passengers via horse-drawn street car between the boat dock and the resort, served as preliminary transportation for guests until a railroad to the island was built ten years later.

What is odd about this report is that there was no follow-up on this discovery. Despite the *Savannah Morning News* being one of the newspapers most willing to report on wild man sightings and other oddities within Georgia and across the country, there were no further articles on giants, skeletons, or anything else regarding the tramway in the following months. It is hard to say if this silence points to a hoax, a misrepresentation, a cover-up, or an oversight. Regardless, it is the first in a string of reports of giant skeletons in Georgia in the late 19th century.

The Etowah (Tumlin) Mounds

Location: Bartow County, Northwest Georgia

Excerpt from *Georgia's Landmarks, Memorials, and Legends, Volume II* (1913) by Lucian Lamar Knight

The following item is copied from one of the old scrap-books of Judge Richard H. Clarke. It reads:

"Several years ago an Indian mound was opened near Cartersville, Ga., by a committee of scientists from Smithsonian. After removing the dirt for some distance a layer of large flag-stones was found, which had evidently been dressed by hand, showing that the men who quarried the rock understood the business. These stones were removed, and in a vault beneath them was found the skeleton of a giant, measuring seven feet and two inches. His hair was coarse and jet black, and hung to the waist, the brow being ornamented with a copper crown. The skeleton was remarkably well preserved and was taken from the vault intact. Nearby were found the bodies of several children of various sizes. The remains of the latter were covered with beads made of bone of some kind. Upon removing these the bodies were found to be enclosed in a network of straw or reeds, and underneath these was a covering of the skin of some animal. In fact, the bodies had been prepared somewhat after the manner of mummies and will doubtless throw new light upon the history of the people who reared these mounds. On the stones which covered the vault were carved inscriptions, and if deciphered will probably lift the veil which has enshrouded the history of the race of giants which undoubtedly at one time inhabited the continent."*

*Extract from a letter written by a Mr. Hazleton to J. B. Toomer and published in the "Banner," of Athens, Ga., date unknown. Reproduced from one of the scrapbooks of Judge Richard H. Clark, in the Carnegie Library, in Atlanta, Ga.

March 4, 1884 - Athens Banner-Watchman (Athens, Clarke County, GA)

Mr. J. B. Toomer yesterday received a letter from Mr. Hazleton, who is on a visit to Cartersville. The letter contained several beads made of bone, and gave an interesting account of the opening of a large Indian mound near that town, by a committee of scientists sent out from the Smithsonian Institute. After removing the dirt for some distance a layer of large flagstones was found, which had evidently been dressed by hand and showed that the men who quarried this rock understood their business. These stones were removed, when in a kind of vault beneath them the skeleton of a giant, that measured 7 feet 2 inches, was found. His hair was coarse and jet black and hung to the waist, the brow being ornamented with a copper crown. The skeleton was remarkably well preserved and taken from the vault intact. Near this skeleton were found the bodies of several children of various sizes. The remains of the latter were covered with beads, made of bone of some kind. Upon removing these, the bodies were found to be enclosed in a network made of straw or reeds, and beneath this was a covering of the skin of some animal. In fact, the bodies had been prepared somewhat after the manner of mummies, and will doubtless throw new light on the history of a people who reared these mounds. Upon the stones that covered the vault were carved inscriptions, and if deciphered will probably lift the veil that has enshrouded the history of the race of giants that undoubtedly at one time inhabited this continent. All the relics were carefully packed and forwarded to the Smithsonian Institute, and are said to be the most interesting collection ever found in America. The explorers are now at work on another mound in Bartow county, and before their return home will visit various sections of Georgia where antiquities are found. On the Oconee river, in Greene county, just above Powell's Mills, are several large mounds, one of them very tall and precipitous. The editor of this paper has now an engagement with Mr. Toon Powell to make an excavation there, and as soon as the weather breaks we will let our readers know what hidden secrets are contained therein.

Excerpt from the "Burial Mounds of the Northern Sections of the United States, by Prof. Cyrus Thomas" Paper within the *Fifth Annual Report of the Bureau of Ethnology to the Secretary of the Smithsonian Institution, 1883-'84* (Published in 1887)

Grave *a*, Fig. 41.—A stone sepulcher, 2 1/2 feet wide, 8 feet long, and 2 feet deep, formed by placing steatite slabs on edge at the sides and ends, and others across the top. The bottom consisted simply of earth hardened by fire. It contained the remains of a single skeleton, lying on its back, with the head east. The frame was heavy and about 7 feet long. The head was resting on a thin copper plate, ornamented with stamped figures; but the skull was crushed and the plate injured by fallen slabs. Under the copper were the remains of a skin of some kind; and under this, coarse matting, probably of a split cane. The skin and matting were both so rotten that they could be secured only in fragments. At the left of the feet were two clay vessels, one a water-bottle, and the other a very small vase. On the right of the feet were some mussel and sea-shells; and immediately under the feet two conch-shells (*Busycon perversum*), partially filled with small shell beads. Around each ankle was a strand of similar beads. The bones and most of the shells were so far decomposed that they could not be saved.

April 13, 1886 - Middle Georgia Argus (Jackson, Butts County, GA)

At Cartersville the water has receded from Tumlin mound-field, and has left uncovered acres of skulls and bones. Some of these are gigantic. If the whole frame is in proportion to two thigh bones that were found, their owner must have stood fourteen feet high. Many curious ornaments of shell, brass and stone have been found. Some of the bodies were buried in small vaults built of stones. The whole makes a mine of archaeological wealth. A representative of the Smithsonian Institute is investigating the curious relics.

Commentary:

The skeleton found in Mound C of the Etowah Mounds—then known as the Tumlin Mounds, after the land owner—in 1884 is the most documented case of a giant skeleton found in Georgia. Not only do we have rumors circulating in the newspapers, but we also have confirmation of a heavy-framed, 7-foot-long skeleton in an official report from the Smithsonian Institute. There, the trail goes cold. Later reports and books on the subject fail to mention that such a skeleton was found. All of the artifacts from the Cyrus Thomas investigations were transported to the Smithsonian, so we can only assume that the skeleton is sitting in a box in a basement in Washington, DC. Perhaps Thomas Steenburg was correct in his interview for the documentary *The Unwonted Sasquatch* when he said, "It wouldn't surprise me if the physical remains of a Sasquatch are laid out on a table one day, they won't be found by a researcher like me in the bush. They'll be found by some anthropology student, who found them in a long-forgotten museum box or drawer somewhere."

As well-documented as the 1884 find was, the 1886 report is completely unfounded. It could have been a hoax, outlandish journalism designed to drive tourist traffic, or simply a giant (pun intended) game of telephone building from 1884 until then. While the 7-foot skeleton of 1884 is undeniable, the 14-foot skeleton of 1886 is a figment of the imagination.

The Scull Shoals Mounds

Location: Greene County, East Georgia

July 9, 1875 - Oglethorpe Echo (Crawford, Oglethorpe County, GA) - Excerpt

Learning that we were near an old Indian burial ground, we expressed a desire to spend the rest of the afternoon examining the same, to which proposition the party readily consented. We landed at a point on the river near where a small stream enters, and after leaving the banks some fifty yards, we reached a spot where a freshet, some years since, had unearthed a cemetery, which, until then, was unknown to any one. The river, in overflowing, washed several large openings in a field, each of which were found filled with human bones and Indian relics. To this day a space of some ten acres is thickly scattered with pieces of their pottery, arrow-heads, numerous strangely shaped stones cut by them, and a quantity of human bones, with which many spots are whitened. They are generally broken into small fragments, but you occasionally find entire skulls and other prominent parts of the human frame. And what is more remarkable, we are told that one often finds the skeleton of a giant, the bones being much larger than those of an ordinary man. Mr. Aycock now has in his possession a peculiarly shaped rock, that he found in one of the graves when they first washed open. It is of oval shape, not quite so large as a breakfast plate, has an indenture on each side like a saucer, is as white as marble, smooth as glass, as regular as if cut by a sculptor, and of pure flint. Two of these rocks were found in a grave, one being red and the other white flint. How the Indian, without any tools, managed to cut this hardest of stone, and for what purpose they were intended, we cannot surmise.

A brisk rain brought to an end our further sport and wanderings for that day, and so we wended our way homeward, where a delightful supper of fish and other etceteras soon caused us to forget our poor luck and "soaking."

Saturday morning we awoke at the break of day, and, on going out, found our party preparing for a sein on Big creek. On this we will not dwell. Suffice it to say that, after a three hour's wade through mud, water and snakes up to our waist, we returned home with a string of fine fish, some four feet long.

After a short rest we again excurted some five miles down the river, to a pair of Indian mounds. After a pleasant ride, in pleasant company, interspersed with an occasional halt in some shady nook to "try our luck," where reached the county line, where Scull Shoals creek separates us from Greene. Just below this point we launched our canoe and went ashore, as the objects of our visit—the mounds—were but a few yards distant. Their towering forms, like huge green balls, were the first object that arrested our sight. They are two in number—a larger and a smaller—and are in a flat formed by a bend in the river, near the banks of the stream. Having no instrument of measurement, we cannot say what was the height or dimensions of either. Of one thing we feel assured—it was no slight undertaking for our party to clamber to the summit of the larger, which could only be accomplished by the aid of the thick bushes on its sides. Of this mound there is only one place at which you can ascend, the other sides being almost perpendicular. Arrived at the summit, we found it covered with a rank growth of tall weeds and bushes. Some years since an enterprising darkey conceived the idea of planting him a water-melon patch on its top, and cleared it off accordingly. We noticed a basin of some thirty feet across in the center, which we suppose was used as a fortification.

Descending, one sees on every hand remnants of Indian pottery, the shells from the mussels they cooked, and even the ashes and charred coals from their fires. All appear just as the owner left it nearly a hundred years ago. This valley is seldom visited, and hence all remains undisturbed. And what is stranger still, human bones cover the earth equally as thick as their pots and arrow heads. In a freshly washed gully we found the entire skeleton of a youth, every bone being in its proper place, while by its side lay several arrow heads, all that were left of the bow and quiver that were laid to rest with the young warrior. Why they made their burial ground beneath their place of abode we cannot conjecture. That they also lived on this spot as well as buried their dead, the pottery, ashes and shells alluded to above clearly confirm.

The smaller mound is not a matter of as much interest as the larger. It stands some 500 yards distant, and not being very steep, is easily ascended.

For what purpose these mounds were formed none can tell. Some historians go so far as to contend that they were not built by the Indian, but by some race that occupied this land prior to them. And this seems plausible, from the fact that the tribes our fathers found here could not tell by whom, when or for what purpose they were built. But we believe them to be the work of the Indian, from the fact that their interiors, when opened, are found to contain the arrow-heads and potteries used only by the red man. It certainly required much time and labor to rear these mighty piles of earth. We can see, near by, the holes from whence the earth was taken to make them, now huge ponds of water.

March 25, 1884 - Athens Banner-Watchman (Athens, Clarke County, GA)

Our chief mission to Greene county was to explore the large Indian mound situated on the left bank of the Oconee, about a mile and a half above Powell's Mills. The aboriginees professed to know nothing of the history of these artificial formations, and it is supposed by some that its construction would date back to the infancy of Uncle Calvin Johnson. Toon Powell had promised to turn over to us his convict force, and calculated that it would take 25 hands three days to explore the largest mound. When we reached his farm Mr. Powell had his wheelbarrows, tools, etc., ready laid out to begin work early on Monday morning. He anticipated no interruption, as the land belonged to some orphan children and Mr. Wray, the tenant gave permission for the exploration. In fact, Mr. W. had already decided to demolish a part of the mound and test its soil as a fertilizer, for it is very black and alluvial, and he had no objection to having a part of the work done for him by other hands. But it seems that we had counted without our host. Sunday evening Mr. Powell received a message from Sheriff Norton, of Greene, who is agent for the property upon which the mound is situated, that he most positively and emphatically objected to the mound being disturbed. Toon had received this message just before we reached his house, and had a bomb exploded at his feet he would not have been more surprised. In fact, in the language of the orator, we found our friend too full for utterance. After gently breaking the melancholy intelligence to us we asked our host what he thought of this unlooked-for quarantine.

"Think! Why my indignation is fatigued. If I had been in Athens when Charlie Norton's message was delivered to me Cran Oliver would have had me in the calaboose before a cat could blink its eye for using forbidden language. To say the least, my thoughts are very deep-seated and emphatic, and would not do to print in a Sunday-school book. But I am determined that you shall not be disappointed. As soon as crops are laid by I will take every hand on my place and throw up a mound that will look like a regular mountain beside a mole-hill when compared with the old pile of dirt up the river, and you may come down next summer and dig into it to your heart's content. I'll show the high sheriff of Greene county that he can't get up any corner in this spread-eagle country of freedom and universal liberty on old second-hand tumuli. I shall build my mound in the most modern and improved style, with bay windows, serpentine walks and an observatory on top. I'll fill it brim full of relics, too, if I have to import them from the banks of the Nile. And if Charlie Norton comes nosing around here while I am at work on it you will find a skeleton, too, in that mound, with a petrified soul about as big as a grain of mustard seed. The truth is, I have had a curiosity for a long time to see what the old mound contained in its inwards, but wanted some other inquisitive jackanapes to make the start and shoulder all responsibility of the expedition. Your proposition to head the explorers came like a ray of sunshine athwart the pathway of my curiosity; but now, right on the heels, as it were, of our lifting a veil that has been drawn for centuries, this veto is hurled at my head. Oh, had I a thousand tongues and was left unconfined in some vast wilderness, that I might do justice to the subject." And Toon sat down a picture of the most abject despair.

We tried to comfort our friend, but it came with poor grace from us, who felt the disappointment about as keenly as did Mr. P. In fact, if anything a little more so, for Toon had to tote the skillet, while our only outlay was loss of time. We asked Mr. Powell if he had a picture of Sheriff Norton, as we wished to engrave his image upon our heart. He said he had not, and what was more his future mission through life would be to blot out all remembrance of that former friend from his mind. "And to think how I loved that man, too!" he wailed, "I asked of him bread and he gave me a stone—and, by jingo, Charlie Norton will find that other people can hurl stones."

This thing of laying up wealth in the shape of old Indian mounds was a new sort of industry to us. We thought if the agency over a little artificial bank of earth would make Sheriff Norton such a nabob, what would be his bearing and sensations if he was elected Captain-General of old Bald Mountain up in Towns county, or owned a section of the Blue Ridge? Why, Georgia wouldn't be large enough to hold that man, and it would be like starting to the end of the rainbow for the common herd to get in speaking distance of him. Vanderbilt would be a regular street beggar beside the sheriff of Greene. We do not know of but one more instance of this kind on record—that of an Oglethorpe farmer who dug a well but continued to use water from his spring, on the ground that the water in his well would keep while that from the spring was wasting. But perhaps this agent thought his old mound was filled with gold and precious stones, and knew full well that the average editor could not bear prosperity and we might die suddenly from enlargement of the pocket after striking this bonanza. As about 2,000 years or more have elapsed since these mounds were built, Mr. Norton doubtless calculated that a nickel deposited in one then, at compound interest, would have swelled to an enormous sum, and he could take the proceeds, pay off the national debt, complete Washington's monument, and thus take a place in the affections of the people alongside of the man who set 'em up to the crowd. Again, perhaps, the High Sheriff of Greene had been reading up on forbidden fruit in the Bible, and interpreted the passage to refer to the little plum orchard on top of the mound, and so felt it his Christian duty to sustain the Scriptures. So it is seen that we extend the greatest of all virtues, Charity, toward Sheriff Norton, and leave him every loophole of escape. But we were determined not to be counted clear out, and so asked Mr. Wray if he thought the agent would consider it a trespass if we went one eye on his old mound. He said he would shoulder all responsibilities if we promised to keep our hands in our pockets and whistle while examining them, for he did not think the Captain would permit any of the precious dirt or broken pottery surrounding them removed. So Monday afternoon, together with Toon Powell, John Davenport, Capt. Frank Pope and Mr. Wray we started for the mounds. Our road led up the river and through a beautiful Bermuda meadow. We progressed very nicely until a ditch in the road was reached, when one of the traces snapped in half. Now navigation in a vehicle without traces is like starting up an engine with the steam turned off. In other words, it can't be did.

The top of the mound could be seen towering in the valley half a mile or more up the river, as if beckoning us to hurry on. Every man in the party turned his pockets inside out, but no signs of a string could be found. Toon made some remark about the devil appearing to be on the side of Charlie Norton, when John Davenport's eye rested on Mr. Wray's watch-guard, made of a little short cloth shoe-string. In an instant he had it off and the trace was repaired. We looked with wonder upon the skill of our young friend, and decided if you would give him a bundle of shoe-strings that he could bind a cyclone hand and foot.

There are two mounds at this place, one very steep and precipitous, there being only one side upon which it can be ascended, while the other is probably twelve or fifteen feet tall and the top levelled off. They are about thirty yards apart. No attempt has ever been made to explore either, although the Harrison freshet washed a good deal of dirt from the base of the larger, exposing a great many relics and a part of the skeleton of a giant. It took only a few minutes to examine the smaller mound, when we turned our attention to the larger, that looked like a huge sugar load rising in the valley. Its sides are covered with a dense growth of bushes and cane, while on its summit is a thrifty plum orchard. A number of years ago Mr. Wray planted a watermelon patch on the top of this mound, and says its soil is exceedingly fertile. Whenever the outer crust is washed away you see pieces of human bones and broken pottery. The relics washed from this mound are superior in finish and design to those made by the Indians. For instance the stone wares are polished as smooth as marble, while the pottery is nicely figured and shows a superior workmanship. If Indians reared these mounds, the race had certainly degenerated greatly at the time of the discovery of America. Dr. J. H. Brightwell, who was raised near this spot, says the largest mound is made of sundried brick, and the clay was not found near at hand, as may be supposed, but brought from different parts of the country for fifty miles or more, as you can see every specimen of soil in this section. The Doctor thinks it was a place of worship, and perhaps required centuries in building, the freezes of winter melting the brick, that were replaced in the spring and more relics deposited. He says that he has conclusive proof that the mound once had an open space or vault on the inside, for he can remember when there was a deep sink on the inside, showing that the supports had given way. Signs of this basin are yet noticeable. Some of the oldest inhabitants say there was a tunnel leading from the top of this mound into the river. This is reasonable, as there is a large mound on the Savannah river, in Elbert county, that had such a passage, and parties were living within the memory of the present generation who had passed through it. But the valleys around these mounds are no less interesting than the elevations themselves. For acres on each side the earth is one vast cemetery, and human bones, relics and broken pottery bestrew the ground. Mr. Wray says that after a rise in the river a few years ago he counted fifteen skeletons exposed by one new-made wash. Mr. Wray secured us a common weeding-hoe, and with it we excavated a number of skeletons. They are not buried in regular order, but the bodies lay in every pasture and position. We found one pasture whose bones had been subjected to fire, for they were charred into coal. Whether the owner of this anatomy had been burned at the stake, or was permitted by his Satanic Majesty to return and warn his people against the world below, is a question that we will not discuss. Mr. Wray says that in plowing this field he finds a great many interesting relics, and his children have gathered long strings of all colored beads from the exposed graves. We are induced to the belief that this spot was the scene of a bloody battle between savage tribes for the possession of the mound, as there are two distinct burial-grounds, with a vacant space between, and the skeletons show that the bodies had been hastily and carelessly interred.

It was a disappointment when forbidden by Mr. Norton to explore these mounds, for it could have done the property no harm whatever and might have thrown new light on the early history of our country. Mr. Wray was exceedingly polite and obliging; but before we separated requested that we permit him to search our pockets, as Mr. Norton would be very angry with him if we had secreted one of his tumuli about our person and carried it off to be explored at leisure. Toon Powell says he intends yet to see what is in the mound, and thinks that after having a talk with Mr. Norton he can prevail upon him to permit the exploration, as he is convinced from his knowledge of that gentleman that he is laboring under some false impression.

Commentary:

As mentioned in the *Athens Banner-Watchman* article on the Etowah Mound C skeleton, the successful excavations at Etowah inspired editor T. Larry Gantt to set his sights on the nearby mounds at Scull Shoals. These mounds are located just south of Athens on the Oconee River in what is now the Oconee National Forest. This was not Mr. Gantt's first experience with the Scull Shoals Mounds, having explored them briefly in 1875 while working as the editor of the *Oglethorpe Echo*. Assisting Gantt was Toon Powell, who—in partnership with his brother-in-law John Davenport—owned the Scull Shoals mill site near the mounds, farmlands in various counties, a railroad contracting company that employed convict labor, and other ventures. Unfortunately for Gantt and Powell, they were prohibited by the landowner from performing a full excavation on the mounds. However, they could walk around the mound complex where the Harrison Freshet, a massive flood in 1840, had damaged and exposed part of the more prominent mound. This article suggests that part of a giant skeleton was found previously, reiterating the rumors heard by Gantt in the 1870s about the finding of giant remains at a location just upriver of the mounds. Still, Gantt and Powell found nothing for themselves.

Nothing like the archaeological work done at Etowah was conducted at Scull Shoals. Some examination and testing of the mounds were undertaken in the 1980s; however, the area could only be minimally excavated. The vegetation within the Oconee National Forest was too thick for much work to be done, and there was evidence of considerable destruction to the site, caused primarily by Mr. Wray's precious watermelon patch and other farming in the 19th century.

Rockmart Cave

Location: Polk County, West Georgia

June 24, 1887 - Hartwell Sun (Hartwell, Hart County, GA)

Three miles east of Rockmart, at the new lime quarry of Crow & Robinson, there was exhumed last week by the workmen a huge specimen of the homo-genus of a prehistoric age.

It is a monster skeleton of a man, measuring two and a half feet across the chest, seven and a half feet in length, feet eighteen inches long, and arms, legs, etc., proportionately long and large. Teeth found near the skeleton, and supposed to belong to it, measure one and a half inches in length.

This skeleton was found in a medium sized cave, the mouth of which was covered with earth eight or ten feet deep, occasioned, it is presumed, by a landslide.

It lay in a large sarcophagus, hewn out by an architect, in the side of the cave, encompassed by great boulders of rock.

Many persons have gone to quarry to see it and examine for themselves. — Rockmart, Ga., Slate

June 29, 1887 - Savannah Morning News (Savannah, Chatham County, GA)

The lime quarry, three miles northeast of Rockmart, continues to be the scene of most remarkable discoveries of animal life in human form of pre-existing ages. Some of the workmen who lived in town had large crowds around them Sunday morning last listening to the wonderful stories they relate relative to the giant skeleton found in a cave ten feet below the surface of the rock. They state that a skeleton of a woman was found lying on its face. When removed or handled carelessly it decomposes. J. H. Dunn visited the quarry last Sunday and says he never saw anything like it. He also told the reporter that some boys found another skeleton on the mountain in some clefts of rocks. He brought a piece of its jawbone along with him, and says "men of this day and time are too small for such jawbones." Another curious feature to be found is small mounds of rock on the floor of the cave, in each of which large human bones are deposited. One of the gentlemen living in town exhibited several pieces of bones from the cave. One piece he brought to town with him has a figure five cut into its surface.—Among many other strange and unaccountable articles found in the cave was a knife of a make and style of age whereunto the memory of man runneth not. The incredulous had better call at the quarry and enlighten their minds.

Commentary:

Polk County is a rich geological area with slate, limestone, iron shale, and clay deposits. Historically, slate has been the most significant export, and the Rockmart Slate Corporation continues to operate a slate quarry that has been mined since the 1850s.

While not typically considered part of Northwest Georgia, Polk County does share geological characteristics with the counties north of it, including an abundance of caves and springs. Archaeological evidence suggests human habitation in the area dates back to at least 12,000 years ago, and the cave formation probably would have predated that. As such, it is impossible to ascertain when these remains would have been laid to rest in the cave.

Based on the description of the various burials within the cave, there is a distinct possibility that these skeletons belonged to Neanderthaloid-type relict hominoids. Neanderthal burials have been found in caves across Europe and western Asia, most notably in France and Iraq. While the exact procedures of Neanderthal funerary practice are still up for debate, they appear to have buried their dead. Many caves have multiple burials that show intentionality in how the bodies were arranged. It also seems that Neanderthals used various techniques for grave formation. Sometimes, they used a natural cave depression as a grave; sometimes, they covered a body with stones; and sometimes, they dug a burial pit. Tools have also been found in these Neanderthal burial caves.

Assuming the reports are accurate, the Rockmart Cave burials seem consistent with Neanderthal burial practices. It is truly a shame that such a find didn't garner any attention from the scientific community.

31

Chesser Island Giant

Location: Chesser Island, Okefenokee Swamp, Charlton County, Southeast Georgia

July 13, 1969 - Atlanta Journal and Constitution (Atlanta, Fulton County, GA) - Excerpt from Article Entitled "Struggle for Okefenokee homestead"

"There come an old professor here from a college way off," Mr. Chesser said. "I spect it was 40 years ago. He hired me to start moving dirt and we uncovered 13 skeletons—he did. He never let me get down to where the people were buried. He uncovered the bones very carefully and it took him a week to do it. Some of the skeletons were crossed, one on top of the other. Some were face down. All of them were perfect when they were first uncovered. Teeth even still had some glaze on them, but when air struck, it crumbled them. They were giants. Those jawbones would go over my whole face.

"The professor took pictures and then hired me to cover everything back up. I wouldn't let them dig until they promised to do that. Since that time, Tom, Dick and Harry wanted to come in, and I wouldn't make them fill up their holes. They ruined it and I'm sorry I ever let them do it.

Commentary:

The unnamed professor who hired Mr. Chesser was most likely the Dr. W. D. Funkhouser mentioned earlier. Dr. Funkhouser became familiar with the Okefenokee area while a junior member of the 1912 Cornell University expedition led by Dr. J. C. Bradley. In 1912, Funkhouser was a Cornell Ph.D. student and high school principal, but he became a college professor after receiving his degree. His initial training was in zoology, and he became head of the University of Kentucky Department of Zoology in 1918. However, zoology was just one of his academic interests. He later became the Graduate School dean and an anthropology professor after founding the Department of Anthropology and Archaeology in 1926 with Dr. W. S. Webb. In 1923, he led an excavation of the Indian Cliff Dwelling at the Pine Mountain School in Harlan County, KY, and Funkhouser and Webb compiled several reports of Indian mounds and other archaeological sites in Kentucky throughout their careers.

Given his familiarity with the Okefenokee and his passion for archaeology, Funkhouser seems the most compelling candidate for the identity of the unknown professor. Unfortunately, though he had some 327 publications to his name, he does not appear to have published anything describing his discovery and analysis of the supposed giant bones at Chesser Island.

Chapter Three: Legends of the Cherokee

Why the Cherokee

Native American legends are vital to studying relict hominoids in North America. This is particularly true in states out west, where written documentation of interactions between white Americans and relict hominoids is limited to the past 150 years, but it is also true in the eastern states. To understand what occurred before European settlement, we must examine the populations predating colonization.

Several indigenous tribes have called Georgia home over its history. The two most prominent are the Cherokee and the Muskogee Creek. Others include the Oconee, the Hitchiti, the Apalachee, the Yamasee, and the Timucua. While all had rich mythological traditions, the Cherokee are the most well-documented. This is partly because of Sequoyah's efforts to develop the Cherokee syllabary in the early 19th century, allowing the oral legends to be preserved in writing in the original language. The syllabary developed by Sequoyah can still be seen today on street signs and business signs in Tahlequah, Oklahoma, the capital city of both the Cherokee Nation and the United Keetoowah Band of Cherokee Indians. We also owe a debt of gratitude to James Mooney, an ethnographer for the US Bureau of American Ethnology who lived with and documented the Cherokees in the late 19th century. His books *The Sacred Formulas of the Cherokees* and *Myths of the Cherokees* form the basis for later scholarship on the Cherokee culture.

The Kechleh-Kudleh

June 27, 1793 - Maryland Gazette (Annapolis, Maryland)

Charleston, May 17.

A gentleman on the South Fork of Saluda river, in a letter of the 23d ult. Sends his correspondent in this city the following description of an extraordinary animal which has been lately discovered on the Bald Mountain, and on other mountains in the western territory:—

This animal is between twelve and fifteen feet high, and in shape resembling a human being, except the head, which is equal in proportion to its body, and draws in somewhat like a terrapin; its feet are like those of a negro, about two feet long, and hairy, which is of a dark dun colour; its eyes are exceedingly large, and open and shut up and down its face; the hair of its head is about six inches long, stands straight like a negro's; its nose is like that of the human species, only large, and inclined to what is called Roman.

These animals are bold, and have lately attempted to kill several persons—in which attempts some of them have been shot.

Their principal resort is on the Bald Mountain, where they lie in wait for travelers—but some have been seen in this part of the country. The inhabitants of this place call it Yahoo; the Indians, however, give it the name of Chickly-Cudly.

Commentary:

This article, which is probably an exaggerated account of an Eastern-Bigfoot-type hominid, is the impetus for the widespread claim that the Cherokee used the term Kecleh-Kudleh, supposedly translated as "hairy man" or "hairy giant," to refer to Bigfoot. The Kecleh-Kudleh story has been perpetuated by many in the field of cryptozoology without validating the research and reasoning themselves. It's an unfortunate example of how supposed indigenous legends often get put on a pedestal of truth without cross-referencing against documented indigenous legends.

When we examine the Cherokee dictionaries available through the work of James Mooney and Dr. Duane King, we find that "Chickly-Cudly" has been neither transliterated nor translated correctly.

Firstly, Cherokee words that have been represented as having a "ch" sound in English have a "ts" sound in Cherokee, not a "k" sound. Examples of this include Cherokee, Chattahoochee, Chattanooga, and so on. Therefore, a word transliterated as "chickly" by 18th-century Southerners was probably spelled "tskili" in the Cherokee language. As it turns out, *tskili'* is a Cherokee word, but it doesn't mean hairy or man. The Cherokee word *tskili'* refers to a witch, a ghost, a night traveler, or a specific type of owl (the dusky great horned owl according to James Mooney's *Myths of the Cherokee* or the long-eared owl according to the *Raven Rock Dictionary*, which is derived from Dr. Duane King's work). *Tskili'* both sounds like "chickly" and has a meaning relevant to the nocturnal characteristics of an Eastern-Bigfoot-type relict hominoid.

This translation makes sense cross-culturally as well. Multiple cultures have associated owls and relict hominoids. For example, in the Bible, in Isaiah 13:21 and 34:14, the se'irim are listed alongside owls and other nocturnal animals, and many researchers, including the esteemed Dmitri Bayanov, feel that the satyrs of Greek mythology and se'irim of Hebrew mythology are rooted in the true history and biology of relict hominoids. The association with owls could be based strictly on their nocturnal nature, but it could also describe a similarity between the howls, hoots, hollers, and other vocalizations of owls and relict hominoids.

The second part of the phrase, Kudleh, is also miswritten and mistranslated. It seems most likely that "cudly" is a variation on *agvdulo*, which means mask, veil, or face covering. . This word is found in the *Raven Rock Dictionary*, in whose phonetic style "v" represents a vowel sound similar to "uh." Loren Coleman describes in his book *The Field Guide to Bigfoot, Yeti, and Other Mystery Primates Worldwide* how in males of the Eastern Bigfoot type—Marked Hominid in his terminology—"the face has hair, or a beard, from the eyes down, so that it looks like they are wearing a mask."

Based on this analysis of the Cherokee language resources available, "masked ghost" or "masked owl-like one" is a far more fitting understanding of what was initially intended by "Chickly-Cudly" than "hairy man" or "hairy giant."

Interestingly, the Cherokee had a tradition of wearing masks while performing dances. Often, these masks were of the animals they would hunt the next day and were intended to bring good luck. However, they also had dances with what has become known as "booger masks," which have an appearance similar to that of a human or a caricature of a human. The Booger Dance was intended to bring protection from evil spirits, and the dance had a much different tone than the dances meant to bring good luck on a hunt. The dancers would don the booger masks and tattered clothing and would chase the women of the tribe, making obscene and suggestive gestures. Anthropologists have categorized this practice as a satire of non-native cultures and even themselves. Still, this dance, like the owl connection, has similarities to mythology based on relict hominoids from other cultures. Both the mythology of the satyrs and the Catholic descriptions of incubi involve hominoid-inspired creatures frequently accosting women. The chasing of women is similar to the practices of the Roman Lupercalia festival in honor of the satyr god Pan, during which young men would dress in animal skins and run through the streets, flogging women for the sake of fertility. It should also not be ignored that the terms "booger," "bogey," and "bugger" all come from the European tradition of the boogeyman, and the word "booger" was explicitly used to describe relict hominoids in American newspapers.

The Stone Man

Nûñ'yunu'wĭ, The Stone Man - Excerpt from James Mooney's *Myths of the Cherokee*

This is what the old men told me when I was a boy.

Once when all of the people of the settlement were out in the mountains on a great hunt one man who had gone one ahead climbed to the top of a high ridge and found a large river on the other side. While he was looking across he saw an old man walking about on the opposite ridge, with a cane that seemed to be made of some bright, shining rock. The hunter watched and saw that every little while the old man would point his cane in a certain direction, then draw it back and smell the end of it. At last he pointed it in the direction of the hunting camp on the other side of the mountain, and this time when he drew back the staff he sniffed it several times as if it smelled very good, and then started along the ridge straight for the camp. He moved very slowly, with the help of the cane, until he reached the end of the ridge, when he threw the cane out into the air and it became a bridge of shining rock stretching across the river. After he had crossed over the bridge it became a cane again, and the old man picked it up and started over the mountain toward the camp.

The hunter was frightened, and felt sure that it meant mischief, so he hurried on down the mountain and took the shortest trail back to the camp to get there before the old man. When he got there and told his story, the medicine-man said the old man was a wicked cannibal monster called Nûñ'yunu'wĭ, "Dressed in Stone," who lived in that part of the county, and was always going about the mountains looking for some hunter to kill and eat. It was very hard to escape from him, because his stick guided him like a dog, and it was nearly as hard to kill him, because his whole body was covered with a skin of solid rock. If he came he would kill and eat them all, and there was only one way to save themselves. He could not bear to look upon a menstrual woman, and if they could find seven menstrual women to stand in the path as he came along the sight would kill him.

So they asked among all the women, and found seven who were sick in that way, and with one of them it had just begun. By the order of the medicine-man they stripped themselves and stood along the path where the old man would come. Soon they heard Nûñ'yunu'wĭ coming through the woods, feeling his way with his stone cane. He came along the trail to where the first woman was standing, and as soon as he saw her he started and cried out: "*Yu!* my grandchild; you are in a very bad state!" He hurried past her, but in a moment he met the next woman, and cried out again: "*Yu!* my child; you are in a terribly way," and hurried past her, but now he was vomiting blood. He hurried on and met the third and the fourth and the fifth woman, but with each one that he saw his step grew weaker until when he came to the last one, with whom the sickness had just begun, the blood poured from his mouth and he fell down on the trail.

The medicine-man drove seven sourwood stakes through his body and pinned him to the ground, and when night came they piled great logs over him and set fire to them, and all the people gathered around to see. Nûñ'yunu'wĭ was a great ada'wehĭ and knew many secrets, and now as the fire came close to him he began to talk, and told them the medicine for all kinds of sickness. At midnight he began to sing, and sang the hunting songs for calling up the bear and the deer and all the animals of the woods and mountains. As the blaze grew hotter his voice sank low and lower, until at last when daylight came, the logs were a heap of white ashes and the voice was still.

Then the medicine-man told them to rake off the ashes, and where the body had lain they found only a large lump of red wâ'dĭ paint and a magic u'lûñsû'ti stone. He kept the stone for himself, and calling the people around him he painted them, on face and breast, with the red wâ'dĭ, and whatever each person prayed for while the painting was being done—whether for hunting success, for working skill, or for a long life—that gift was his.

Definitions from James Mooney's *Myths of the Cherokee*:

Ada'wehĭ — a magician or supernatural being

Nûñ'yunu'wĭ — contracted from *Nûñyû-unu'wĭ*. "Stone-clad," from *nûñyû*, rock, and *agawănu'wú*, "I am clothed or covered." A mythic monster, invulnerable by reason of his stony skin.

U'lûñsû'ti — "Transparent"; the great talismanic crystal of the Cherokee.

Wâ'dĭ — paint, especially red paint.

Commentary:

Many in the cryptozoological community have referenced the Stone Man as one of the Cherokee names for relict hominoids; however, in reading the entire legend, it is clear that there are other aspects to the Stone Man archetype. Where we should expect to see hints of a biological origin to the story, we instead see every indication of a cosmological origin.

For one, the use of women experiencing their menses to conquer the Stone Man is inconsistent with accounts of relict hominoids. Instead, cryptozoological, paranormal, and preternatural events increase when women are experiencing their menses. This increase is documented in modern reports and folklore related to faeries and other non-human entities.

Additionally, the repeated use of the number seven is a pretty immediate indication that we should look to the skies, as these would be analogous to the seven classical planets or the seven sisters of the Pleiades.

Finally, the Stone Man mimics the Ophiuchus/Asclepius archetype: being knowledgeable in healing, carrying a staff, and being associated with a great fire. In the sky, this great fire is the brightest area of the Milky Way, which is next to the Ophiuchus constellation.

The Bear Man

The Bear Man - Excerpt from James Mooney's *Myths of the Cherokee*

A man went hunting in the mountains and came across a black bear, which he wounded with an arrow. The bear turned and started to run the other way, and the hunter followed, shooting one arrow after another into it without bringing it down. Now, this was a medicine bear, and could talk or read the thoughts of people without their saying a word. At last he stopped and pulled the arrows out of his side and gave them to the man, saying, "It is of no use for you to shoot at me, for you can not kill me. Come to my house and let us live together." The hunter thought to himself, "He may kill me;" but the bear read his thoughts and said, "No, I won't hurt you." The man thought again, "How can I get anything to eat?" but the bear knew his thoughts, and said, "There shall be plenty." So the hunter went with the bear.

They went on together until they came to a hole in the side of the mountain, and the bear said, "This is not where I live, but there is going to be a council here and we will see what they do." They went in, and the hole widened as they went, until they came to a large cave like a townhouse. It was full of bears—old bears, young bears, and cubs, white bears, black bears, and brown bears—and a large white bear was the chief. They sat down in a corner, but soon the bears scented the hunter and began to ask, "What is it that smells bad?" The chief said, "Don't talk so; it is only a stranger come to see us. Let him alone." Food was getting scarce in the mountains, and the council was to decide what to do about it. They had sent out messengers all over, and while they were talking two bears came in and reported that they had found a country in the low grounds where there were so many chestnuts and acorns that mast was knee deep. Then they were all pleased, and got ready for a dance, and the dance leader was the one the Indians call Kalâs'-gûnăhi'ta, "Long Hams," a great black bear that is always lean. After the dance the bears noticed the hunter's bow and arrows, and one said, "This is what men use to kill us. Let us see if we can manage them, and may be we can fight man with his own weapons." So they took the bow and arrows from the hunter to try them. They fitted the arrow and drew back the string, but when they let go it caught in their long claws and the arrows dropped to the ground. They saw that they could not use the bow and arrows and gave them back to the man. When the dance and the council were over, they began to go home, excepting the White Bear chief, who lived there, and at last the hunter and the bear went out together.

They went on until they came to another hole in the side of the mountain, when the bear said, "This is where I live," and they went in. By this time the hunter was very hungry and was wondering how he could get something to eat. The other knew his thoughts, and sitting up on his hind legs he rubbed his stomach with his forepaws—*so*—and at once he had both paws full of chestnuts and gave them to the man. He rubbed again—*so*—and gave the man both paws full of blackberries. He rubbed again—*so*—and had his paws full of acorns, but the man said that he could not eat them, and that he had enough already.

The hunter lived in the cave with the bear all winter, until long hair like that of a bear began to grow all over his body and he began to act like a bear; but he still walked like a man. One day in early spring the bear said to him, "Your people down in the settlement are getting ready for a grand hunt in these mountains, and they will come to this cave and kill me and take these clothes from me"—he meant his skin—"but they will not hurt you and will take you home with them." The bear knew what the people were doing down in the settlements just as he always knew what the man was thinking about. Some days passed and the bear said again, "This is the day when the Topknots will come to kill me, but the Split-noses will come first and find us. When they have killed me they will drag me outside the cave and take off my clothes and cut me in pieces. You must cover the blood with leaves, and when they are taking you away look back after you have gone a piece and you will see something."

Soon they heard the hunters coming up the mountain, and then the dogs found the cave and began to bark. The hunters came and looked inside and saw the bear and killed him with their arrows. Then they dragged him outside the cave and skinned the body and cut it in quarters to carry home. The dogs kept barking until the hunters thought there must be another bear in the cave. They looked in again and saw the man at the farther end. At first they thought it was another bear on account of his long hair, but they soon saw it was the hunter who had been lost the year before, so they went in and brought him out. Then each hunter took a load of the bear meat and they started home again, bringing the man and the skin with them. Before they left the man piled leaves over the spot where they had cut up the bear, and when they had gone a little way he looked behind and saw the bear rise up out of the leaves, shake himself, and go back into the woods.

When they came near the settlement the man told the hunters that he must be shut up where no one could see him, without anything to eat or drink for seven days and nights, until the bear nature had left him and he became like a man again. So they shut him up alone in a house and tried to keep very still about it, but the news got out and his wife heard of it. She came for her husband, but the people would not let her near him; but she came every day and begged so hard that at last after four or five days they let her have him. She took him home with her, but in a short time he died, because he still had a bear's nature and could not live like a man. If they had kept him shut up and fasting until the end of the seven days he would have become a man again and would have lived.

Definitions from James Mooney's *Myths of the Cherokee*:

Kalâs'-gûnăhi'ta — "long hams' (*gûnăhi'ta*, "long"); a variety of bear.

Commentary:

The story of the Bear Man is rarely discussed in cryptozoological circles, but it is far more compelling than the Stone Man legend as it introduces the *kalâs'-gûnăhi'ta*.

According to another portion of *Myths of the Cherokee*, the *kalâs'-gûnăhi'ta* is "a large black bear with long legs and small feet, which is always lean, and which the hunter does not care to shoot, possibly on account of its leanness." The passage also mentions that "it is believed that new-born cubs are hairless, like mice." A relict hominoid could resemble a long, lean bear, and it is reasonable to infer that relict hominoid infants could appear hairless at birth, with the dense coat of body hair not fully developing until puberty, similar to the less extensive body hair of modern humans.

The tale's main character also developed the characteristics of a bear while living in the woods for quite some time. Despite continuing to walk on two legs, he grew a full coat of body hair and more animal-like behavior. Throughout this book, from the Athabaskan lore of the Bushman to the newspaper accounts of Neanderthaloids in the hills of North and Central Georgia, we see a repeated belief that modern man can grow body hair and lose his humanity simply by living in the wild for an extended period. The Cherokee legend of the Bear Man repeats this motif, suggesting they, too, were familiar with Neanderthaloid-type relict hominoids.

Yawâ'ï

Yawâ'ï - Excerpt from James Mooney's *Myths of the Cherokee*

"Yawa place," a spot on the south side of Yellow creek of Cheowa river, in Graham county, about a mile above the trail crossing near the mouth of the creek. The legend is that a mysterious personage, apparently a human being, formerly haunted a round knob near there, and was sometimes seen walking about the top of the knob crying, *Yawă'! Yawă'!* while the sound of invisible guns came from the hill, so that the people were afraid to go near it.

Commentary:

While this is technically a local legend of North Carolina, the Cherokee were prolific throughout the southern Appalachian mountains. Graham County is in the southwest corner of North Carolina, on the border with Tennessee and just one county north of the border with Georgia. The Cherokee in the Georgia settlements would have been familiar with this legend.

The "*Yawă'! Yawă'!*" cry is similar to some vocalizations associated with relict hominoids, and the "invisible guns" is likely an explanation of what is now referred to as wood knocks in the cryptozoological community.

The Slant-Eyed Giant

Tsul'kălû', The Slant-Eyed Giant - Excerpt from James Mooney's *Myths of the Cherokee*

A long time ago a widow lived with her one daughter at the old town of Kănuga on Pigeon river. The girl was of age to marry, and her mother used to talk with her a good deal, and tell her she must take no one but a good hunter for a husband, so that they would have plenty of meat in the house. The girl said such a man was hard to find, but her mother advised her not to be in a hurry, and to wait until the right one came.

Now the mother slept in the house while the girl slept outside in the âsĭ. One dark night a stranger came to the âsĭ wanting to court the girl, but she told him her mother would let her marry no one but a good hunter. "Well," said the stranger, "I am a great hunter," so she let him come in, and he stayed all night. Just before day he said he must go back now to his own place, but that he had brought some meat for her mother, and she would find it outside. Then he went away and the girl had not seen him. When day came she went out and found there a deer, which she brought into the house to her mother, and told her it was a present from her new sweetheart. Her mother was pleased, and they had deersteaks for breakfast.

He came again the next night, but again went away before daylight, and this time he left two deer outside. The mother was more pleased this time, but said to her daughter, "I wish your sweetheart would bring us some wood." Now wherever he might be, the stranger knew their thoughts, so when he came the next time he said to the girl, "Tell your mother I have brought the wood"; and when she looked out in the morning there were several great trees lying in front of the door, roots and branches and all. The old woman was angry, and said, "He might have brought us some wood that we could use instead of whole trees that we can't split, to litter up the road with brush." The hunter knew what she said, and the next time he came he brought nothing, and when they looked out in the morning the trees were gone and there was no wood at all, so the old woman had to go after some herself.

Almost every night he came to see the girl, and each time he brought a deer or some other game, but still he always left before daylight. At last her mother said to her, "Your husband always leaves before daylight. Why don't he wait? I want to see what kind of a son-in-law I have." When the girl told this to her husband he said he could not let the old woman see him, because the sight would frighten her. "She wants to see you, anyhow," said the girl, and began to cry, until at last he had to consent, but warned her that her mother must not say that he looked frightful (*usga'sě′ti'yu*).

The next morning he did not leave so early, but stayed in the âsĭ, and when it was daylight the girl went out and told her mother. The old woman came and looked in, and there she saw a great giant, with long slanting eyes (*tsul′kălû′*), lying doubled up on the floor, with his head against the rafters in the left-hand corner at the back, and his toes scraping the roof in the right-hand corner by the door. She gave only one look and ran back to the house, crying, *Usga'sě′ti'yu! Usga'sě′ti'yu!*

Tsul′kălû′ was terribly angry. He untwisted himself and came out of the âsĭ, and said good-bye to the girl, telling her that he would never let her mother see him again, but would go back to his own country. Then he went off in the direction of Tsunegûñ′yĭ.

Soon after he left the girl had her monthly period. There was a very great flow of blood, and the mother threw it all into the river. One night after the girl had gone to bed in the âsĭ her husband came again to the door and said to her, "It seems you are alone," and asked where was the child. She said there had been none. Then he asked where was the blood, and she said that her mother had thrown it into the river. She told just where the place was, and he went there and found a small worm in the water. He took it up and carried it back to the âsĭ, and as he walked it took form and began to grow, until, when he reached the âsĭ, it was a baby girl that he was carrying. He gave it to his wife and said, "Your mother does not like me and abuses our child, so come and let us go to my home." The girl wanted to be with her husband, so, after telling her mother good-bye, she took up the child and they went off together to Tsunegûñ′yĭ.

Now, the girl had an older brother, who lived with his own wife in another settlement, and when he heard that his sister was married he came to pay a visit to her and her new husband, but when he arrived at Kănuga his mother told him his sister had taken her child and gone away with her husband, nobody knew where. He was sorry to see his mother so lonely, so he said he would go after his sister and try to find her and bring her back. It was easy to follow the footprints of the giant, and the young man went along the trail until he came to a place where they had rested, and there were tracks on the ground where a child had been lying and other marks as if a baby had been born there. He went on along the trail and came to another place where they had rested, and there were tracks of a baby crawling about and another lying on the ground. He went on and came to where they had rested again, and there were tracks of a child walking and another crawling about. He went on until he came where they had rested again, and there were tracks of one child running and another walking. Still he followed the trail along the stream into the mountains, and came to the place where they had rested again, and this time there were footprints of two children running all about, and the footprints can still be seen in the rock at that place.

Twice again he found where they had rested, and then the trail led up the slope of Tsunegûñ'yĭ, and he heard the sound of a drum and voices, as if people were dancing inside the mountain. Soon he came to a cave like a doorway in the side of the mountain, but the rock was so steep and smooth that he could not climb up to it, but could only just look over the edge and see the heads and shoulders of a great many people dancing inside. He saw his sister dancing among them and called to her to come out. She turned when she heard his voice, and as soon as the drumming stopped for a while she came out to him, finding no trouble to climb down the rock, and leading her two little children by the hand. She was very glad to meet her brother and talked with him a long time, but did not ask him to come inside, and at last he went away without having seen her husband.

Several other times her brother came to the mountain, but always his sister met him outside, and he could never see her husband. After four years had passed she came one day to her mother's house and said her husband had been hunting in the woods near by, and they were getting ready to start home to-morrow, and if her mother and brother would come early in the morning they could see her husband. If they came too late for that, she said, they would find plenty of meat to take home. She went back into the woods, and the mother ran to tell her son. They came to the place early the next morning, but Tsul'kălû' and his family were already gone. One the drying poles they found the bodies of freshly killed deer hanging, as the girl has promised, and there were so many that they went back and told all their friends to come for them, and there were enough for the whole settlement.

Still the brother wanted to see his sister and her husband, so he went again to the mountain, and she came out to meet him. He asked to see her husband, and this time she told him to come inside with her. They went in as through a doorway, and inside he found it like a great townhouse. They seemed to be alone, but his sister called aloud, "He wants to see you," and from the air came a voice, "You can not see me until you put on a new dress, and then you can see me." "I am willing," said the young man, speaking to the unseen spirit, and from the air came the voice again, "Go back, then, and tell your people that to see me they must go into the townhouse and fast seven days, and in all that time they must not come out from the townhouse or raise the war whoop, and on the seventh day I shall come with new dresses for you to put on so that you can all see me."

The young man went back to Kănuga and told the people. They all wanted to see Tsul'kălû', who owned all the game in the mountains, so they went into the townhouse and began the fast. They fasted the first day and the second and every day until the seventh—all but one man from another settlement, who slipped out every night when it was dark to get something to eat and slipped in again when no one was watching. On the morning of the seventh day the sun was just coming up in the east when they heard a great noise like the thunder of rocks rolling down the side of Tsunegûñ'yĭ. They were frightened and drew near together in the townhouse, and no one whispered. Nearer and louder came the sound until it grew into an awful roar, and every one trembled and held his breath—all but one man, the stranger from the other settlement, who lost his senses from fear and ran out of the townhouse and shouted the war cry.

At once the roar stopped and for some time there was silence. Then they heard it again, but as if it were going farther away, and then farther and farther, until at last it died away in the direction of Tsunegûñ'yĭ, and then all was still again. The people came out from the townhouse, but there was silence, and they could see nothing but what had been seven days before.

Still the brother was not disheartened, but came again to see his sister, and she brought him into the mountain. He asked why Tsul'kălû' had not brought the new dresses, as he had promised, and the voice from the air said, "I came with them, but you did not obey my word, but broke the fast and raised the war cry." The young man answered, "It was not done by our people, but by a stranger. If you will come again, we will surely do as you say." But the voice answered, "Now you can never see me." Then the young man could not say any more, and he went back to Kănuga.

Definitions from James Mooney's *Myths of the Cherokee*:

Âsĭ — the sweat lodge and occasional winter sleeping apartment of the Cherokee and other southern tribes. It was a low-built structure of logs covered with earth, and from its closeness and the fire usually kept smoldering within was known to the old traders as the "hot house."

Kănuga — a lower Cherokee settlement. The name signifies "a scratcher," a sort of bone-toothed comb with which ball-players are scratched upon their naked skin preliminary to applying the conjuring medicine; *de'tsinuga'skú*, "I am scratching it."

Tsul'kălû' — "Slanting-eyes," literally "He had them slanting" (or leaning up against something); the prefix *ts* makes it a plural form, and the name is understood to refer to the eyes, although the word eye (*aktă*, plural *diktă*) is not a part of it.

Tsunegûñ'yĭ — Tennessee bald, at the extreme head of Tuckasegee river, on the east line of Jackson county, North Carolina. The name seems to mean, "There where it is white," from *ts*, a prefix indicating distance, *une'gă*, white, and *yĭ*, locative.

Usga'sĕ'ti'yu — very dangerous, very terrible; intensive of *usga'sĕ'ti*.

Commentary:

There are elements to this story that are pretty unbelievable and, similar to the Stone Man legend, suggest a cosmological interpretation is the correct one. For example, anyone who has seen how babies are born knows that it is patently ridiculous from a biological perspective for a baby to grow from a worm expelled as part of a woman's menses; thus, that aspect of the story probably was not intended to be interpreted biologically.

There is also the fact that the brother found where the giant and his wife rested seven times along their journey, and the family was asked to fast for seven days to see the giant. Again, we should immediately be skeptical of a literal interpretation when the number seven is repeatedly involved.

However, given that the following Cherokee legend indicates there may be some historical context for giants like *Tsul'kălû'*, we should give his story consideration beyond the cosmological.

One thing to consider is the name of *Tsul'kălû'* and its translation. The Cherokee used the word *Tsul'kălû'* prolifically in place names, so through both geography and mythology, it has had many variations and bastardizations over the centuries: Tuli-cula, Jutaculla, Judaculla, and others. *Tsul'kălû'* is understood by James Mooney and others as referring to slanting eyes, but it is not a literal translation as the word *Tsul'kălû'* does not contain the Cherokee word for eyes. Historical scholars have suggested the meaning may derive from the particular manner in which Plains tribes plucked their eyebrows out and wore their hair, giving a strange appearance to the eyes. Bigfoot researchers and enthusiasts have suggested that it refers to the heavy brow ridge of relict hominoids, which could produce the impression of slanting eyes relative to a human face.

First, though, we should attempt to ascertain if a reference to the eyes should be understood in *Tsul'kălû'*. Three similar words that James Mooney cross-references in his glossary are *Ătă'-gûl'kălû'*, *Gûl'kă-la'skĭ*, and *Tsundige'wĭ*. *Ătă'-gûl'kălû'* was an 18th-century Cherokee chief whose name meant "leaning wood." *Ătă'* is the Cherokee word for wood, and *gûl'kălû'* is the same verb in the word *Tsul'kălû'* that means "something long is leaning, without sufficient support, against some other object." *Ătă'-gûl'kălû'* was described by natural historian and explorer William Bartram as quite frail in appearance, particularly as compared to the other Cherokee, so it is supposed that his name derived from his frail nature and his need for support, perhaps in the form of a cane. If we take that exact verb translation and apply it to *Tsul'kălû'*, with *ts* implying plurality, then *Tsul'kălû'* should refer to multiple things leaning and in need of support. Similarly, *Gûl'kă-la'skĭ* was the name of the Cherokee chief Tsunu'lauhn'ski in his early life, and James Mooney indicates the name means "something habitually falling from a leaning position."

The assumption that *Tsul'kălû'* refers to the eyes is further complicated by the word *Tsundige'wĭ*. The *Tsundige'wĭ* were tiny man-like beings in Cherokee mythology who lived in nests and were constantly plagued by bird attacks. Like *Tsul'kălû'*, the word *Tsundige'wĭ* has an assumed object. It means "they have them closed" and derives from *dige'wĭ*, the plural form of *ge'wĭ*, which means "closed, stopped up, blind." However, the assumed object of *Tsundige'wĭ* is not eyes like *Tsul'kălû'*. Instead, the assumed object is anuses. *Tsundige'wĭ* supposedly translates to "closed anuses," which is an even stranger translation than slanting eyes and is utterly irrelevant to the narrative of the Cherokee teaching the *Tsundige'wĭ* to protect themselves from birds. How do we trust the assumed object of these words, as described by James Mooney, when they seem so outlandish and unrelated to the mythology? Perhaps we shouldn't.

In John Haywood's 1823 treatise, *The Natural and Aboriginal History of Tennessee*, Tuli-cula is an invisible person, not a giant. We also have versions of the *Tsul'kălû'* tale that have been handed down to Jack and Anna Kilpatrick in their work with the Oklahoma Cherokees, those who descended from the survivors of the Trail of Tears. The first story mentions that the Tsuhl'gûl' was so tall that "he would lean on something" and that "he used to fall over upon people and smash them." The second story is more similar to the James Mooney version and indicates that the Tsuhl'gûl' had slanting eyes, a whooping call, and a penchant for whiskey and women.

It seems quite possible that the literal translation of leaning in the name *Tsul'kălû'/*Tsuhl'gûl' initially derived from a leaning, stooped posture but was conflated with slanting eyes as the Cherokee lost touch with their traditions. This interpretation is consistent with the Eastern Bigfoot profile. In her book *Bigfoot in Maine*, Michelle Y. Souliere documents a pattern seen in reports from Maine as "a light bend to the knees and/or a slight tilt forward of the upper body, paraphrasing from one person's description of their stance as kind of leaning forward but not off-balance."

The alternative version documented by the Kilpatricks illustrate why we should be careful in naively relying on the translations given by James Mooney. While Mooney represents one of the best resources for studying the Cherokee, documenting plants, animals, language, culture, history, and religion, he certainly was not infallible, nor was he a Native speaker of the language.

The Giants from the West

The Giants from the West - Excerpt from James Mooney's *Myths of the Cherokee*

James Wafford, of the western Cherokee, who was born in Georgia in 1806, says that his grandmother, who must have been born about the middle of the last century, told him that she had heard from the old people that long before her time a party of giants had come once to visit the Cherokee. They were nearly twice as tall as common men, and had their eyes set slanting in their heads, so that the Cherokee called them Tsunil'kălû', "The Slant-eyed people," because they looked like the giant hunter Tsul'kălû'. They said that these giants lived very far away in the direction in which the sun goes down. The Cherokee received them as friends, and they some time, and then returned to their home in the west. The story may be a distorted historical tradition.

Definitions from James Mooney's *Myths of the Cherokee*:

Tsul'kălû' — See previous legend.

Tsunil'kălû' — the plural form for *Tsul'kălû'*; a traditional giant tribe in the west.

Commentary:

While this narrative is related to the full *Tsul'kălû'* story just discussed, it takes the giants out of mythology and puts them squarely in the realm of historical interaction.

Scholars, including James Mooney, have debated whether this story is a hold-over from when the Cherokee were still living among their fellow Iroquois in the northeastern United States or if this story would have occurred after their immigration southward. The presumption by historical scholars is that the giants referred to would have been Native Americans from one of the Plains tribes or the Osage tribe of Arkansas and Missouri.

Regardless, the fact that a prehistoric interaction between an entire tribe of giants and the Cherokee or their ancestors indicates that the Tsul'kălû' legend may not strictly be mythological.

The Little People

The Little People - Excerpt from James Mooney's *Myths of the Cherokee*

There is another race of spirits, the *Yŭñwĭ Tsundi'*, or "Little People," who live in rock caves on the mountain side. They are little fellows, hardly reaching up to a man's knee, but well shaped and handsome, with long hair falling almost to the ground. They are great wonder workers and are very fond of music, spending half of their time drumming and dancing. They are helpful and kind-hearted, and often when people have been lost in the mountains, especially children who have strayed away from their parents, the Yŭñwĭ Tsundi' have found them and taken care of them and brought them back to their homes. Sometimes their drum is heard in lonely places in the mountains, but it is not safe to follow it, because the Little People do not like to be disturbed at home, and they throw a spell over the stranger so that he is bewildered and loses his way, and even if he does at last get back to the settlement he is like one dazed ever after. Sometimes, also, they come near a house at night and the people inside hear them talking, but they must not go out, and in the morning they find the corn gathered or the field cleared as if a whole force of men had been at work. If anyone should go out to watch, he would die. When a hunter finds anything in the woods, such as a knife or a trinket, he must say, "Little People, I want to take this," because it may belong to them, and if he does not ask their permission they will throw stones at him as he goes home.

Once a hunter in winter found tracks in the snow like the tracks of little children. He wondered how they could have come there and followed them until they led him to a cave, which was full of Little People, young and old, men, women, and children. They brought him in and were kind to him, and he was with them some time; but when he left they warned him that he must not tell or he would die. He went back to the settlement and his friends were all anxious to know where he had been. For a long time he refused to say, until at last he could not hold out any longer, but told the story, and in a few days he died. Only a few years ago two hunters from Raventown, going behind the high fall near the head of Oconaluftee on the East Cherokee reservation, found there a cave with fresh footprints of the Little People all over the floor.

During the smallpox among the East Cherokee just after the war one sick man wandered off, and his friends searched, but could not find him. After several weeks he came back and said that the Little People had found him and taken him to one of their caves and tended him until he was cured.

About twenty-five years ago a man named Tsantăwû' was lost in the mountains on the head of Oconaluftee. It was winter time and very cold and his friends thought he must be dead, but after sixteen days he came back and said that the Little People had found him and taken him to their cave, where he had been well treated, and given plenty of everything to eat except bread. This was in large loaves, but when he took them in his hand to eat they seemed to shrink into small cakes so light and crumbly that though he might eat all day he would not be satisfied. After he was well rested they had brought him a part of the way home until they came to a small creek, about knee deep, when they told him to wade across to reach the main trail on the other side. He waded across and turned to look back, but the Little People were gone and the creek was a deep river. When he reached home his legs were frozen to the knees and he lived only a few days.

Definitions from James Mooney's *Myths of the Cherokee*:

Yûñwĭ Tsundi' - "Little People," from *yûñwĭ*, person, people, and *tsunsdi'gă* or *tsunsdi'*, plural of *usdi'gă* or *usdi'*, little; the Cherokee fairies.

Commentary:

The Cherokee legend of the *Yûñwĭ Tsundi'*, or Little People, is inconclusive as proof of the existence of relict hominoids. The stories have a heavily mythological element, evidenced by the fact that the *Yûñwĭ Tsundi'* are known as the Cherokee fairies. Still, there may be truth behind the mythology, with Proto-Pygmy-type relict hominoids serving as a biological framework for the legend.

The tiny footprints and long, flowing hair of the *Yûñwĭ Tsundi'* match that of the Proto-Pygmies. Also compelling is that the *Yûñwĭ Tsundi'* live in nests built in trees, as the Proto-Pygmies are reportedly adept at tree climbing. On the other hand, the legend differs significantly from biology in the size of the Yûñwĭ Tsundi', which are said to be only as tall as a man's knee, much more petite than even the Proto-Pygmies.

Unfortunately, with so many similarities to Celtic fairy lore, such as the tendency of witnesses to die shortly after their encounter, it is difficult to say how much of the legend is derived from biology, how much is derived from a strictly Cherokee source, and how much may have been assimilated from the tales of the Scottish and Irish immigrants to the Appalachian region.

Chapter Four: Legends of the Okefenokee

The Land of the Trembling Earth

The Okefenokee Swamp is a vast and mysterious wetland on the border between Georgia and Florida. It is the largest blackwater swamp in North America, encompassing 700 square miles and home to alligators, black bears, and a cornucopia of other wildlife. It was considered relatively impenetrable for many years because of the immense effort required to move about. Eighteenth-century explorer Charles Floyd called it the "great monster." Traditionally, Native Americans had used the swamp to hide from other native tribes, the Spanish, and the English. Eventually, Georgia cracker communities came to thrive on the edges of the swamp, but throughout this period, the heart of the swamp maintained its secrets.

However, with the advent of steam power technology, various corporations built railroads into the swamp to harvest abundant old-growth pine and cypress. Plans for an agricultural community would have transformed the land even further had attempts to drain the swamp not failed so dramatically. It wasn't until 1937 that the Okefenokee was purchased by the federal government and designated as a wildlife refuge. Even so, this was a controversial decision. It preserved the habitat and wildlife from corporate desecration, but it eliminated the ability of the local population to survive in the harsh environment and effectively ended Georgia cracker culture in the Okefenokee.

Given the wildness of the land, you may assume that the Okefenokee would be a veritable buffet of wild man and relict hominoid tales, but the opposite is true. While it does have some legends, the Okefenokee never had enough interaction between human and hominoid populations for a significant corpus of witness testimony to develop. Before the logging industry took over, the Okefenokee could have easily housed relict hominoids in its interior while the human populations on the perimeter remained blissfully unaware. After the logging industry took over, there was so much habitat destruction that the local relict hominoid population likely relocated or died out. There is evidence of a modern population in the Okefenokee, but the swamp has returned to a wild state with few humans in the area to serve as witnesses.

The tales that persist from the pre-1937 era seem more like legends than witness testimony, but they have been included for cultural context nonetheless.

Excerpt from Travels Through North and South Carolina, Georgia, East and West Florida, the Cherokee Country, the Extensive Territories of the Muscogulges or Creek Confederacy, and the Country of the Chactaws (1792) by William Bartram

The river St. Mary has its source from a vast lake, or marsh, call Okefenokee, which lies between Flint and Ocmulgee rivers, and occupies a space of near three hundred miles in circuit. This vast accumulation of waters, in the wet season, appears as a lake, and contains some large islands or knolls, of rich high land; one of which the present generation of the Creeks represent to be a most blissful spot of the earth: they say it is inhabited by a peculiar race of Indians, whose women are incomparably beautiful; they also tell you that this terrestrial paradise has been seen by some of their enterprising hunters, when in pursuit of game, who being lost in inextricable swamps and bogs, and on the point of perishing, were unexpectedly relieved by a company of beautiful women, whom they call daughters of the sun, who kindly gave them such provisions as they had with them, which were chiefly fruit, oranges, dates, &c. and some corn cakes, and then enjoined them to fly for safety to their own country; for that their husbands were fierce men, and cruel to strangers: they further say, that these hunters had a view of their settlements, situated on the elevated banks of an island, or promontory, in a beautiful lake; but that in their endeavors to approach it, they were involved in perpetual labyrinths, and, like enchanted land, still as they imagined they had just gained it, it seemed to fly before them, alternately appearing and disappearing. They resolved, at length, to leave the delusive pursuit, and to return; which, after a number of inexpressible difficulties, they effected. When they reported their adventures to their countrymen, their young warriors were enflamed with an irresistible desire to invade, and make a conquest of, so charming a country; but all their attempts hitherto have proved abortive, never having been able again to find that enchanting spot, nor even any road or pathway to it; yet they say that they frequently meet with certain signs of its being inhabited, as the building of canoes, footsteps of men, &c. They tell another story concerning the inhabitants of the sequestered country, which seems probable enough, which is, that they are the posterity of a fugitive remnant of the ancient Yamasees, who escaped massacre after a bloody and decisive conflict between them and the Creek nation (who, it is certain, conquered, and nearly exterminated, that once powerful people), and here found an asylum, remote and secure from the fury of their proud conquerers. It is, however, certain that there is a vast lake, or drowned swamp, well known, and often visited both by white and Indian hunters, and on its environs the most valuable hunting grounds in Florida, well worth contending for, by those powers whose territories border upon it. From this great source of rivers, St. Mary arises, and meanders through a vast plain and pine forest, near an hundred and fifty miles to the ocean, with which it communicates, between the points of the Amelia and Talbert islands; the waters flow deep and gently down from its source to the sea.

December 4, 1921 - Atlanta Constitution (Atlanta, GA)

Still within that shadowy realm of mystery, and probably fated to remain a mere fascinating legend, is the "Land of The Daughters of The Sun," a land warmed almost into reality in the hearts of the people who claim the great Okefenokee swamp as their home and who have heard the quaint legend from the tongues of passing generations.

In the rich store of lore and legend that throws a glamour on intense romanticism about America's most primitive haunt of wild life and untrammeled nature, no story is more thoroughly charming than that of the "Land of The Daughters of The Sun."

Soon after I passed into the rim of the mighty Okefenokee I heard the tale. I am told that historians who have written briefly upon the Okefenokee—and there are none who have devoted extensive work or space to the immense natural wonderland—have mentioned the story and perhaps commented upon its origin. However, its charm fits it for repetition.

Before the earliest Seminoles had sought a retreat in the heart of the great marsh, according to the legend, there lived on an island in its very heart a race of people whose women were so beautiful that to look upon them was to become incurably enamored of them. Of such ravishing beauty and grace were these copper-colored goddesses, the story says, that the race became known as the "People of The Daughters of The Sun," and the island on which they lived was named the "Land of The Daughters of The Sun."

When the first adventurous band of Seminoles penetrated the swamp, it came upon the enchanted island, and the exploring braves of the outer world immediately succumbed to the rich blushes that mounted to the cheeks of the "Sun Daughters" and the raven luster of their hair and eyes.

The legend tells that when the Seminoles begged to be allowed to enter the island, the "Sun Daughters" refused, warning them that their own braves were soon to return from fishing and hunting and would slay the newcomers at sight.

Feigning acquiescence, the Seminoles left for the rim of the swamp, but the object of their return was to secure heavy reinforcements and to make their way back to the island and slay the men of the legendary tribe by stealth and craft, in order that they might win the "Daughters of the Sun" to be their wives.

It was a fierce and mighty war party that silently worked its way into the swamp a second time, it is said, and when the painted and feather-bedecked warriors neared the mysterious island, the utmost stealth and skill of the Indian was employed to keep the coming attack a complete surprise. But as the canoes glided along without the faintest of ripples, a piercing scream that was half human and half animal echoed and re-echoed from above, and a few moments later the light canoes leaped and danced as a great tidal wave swept over the placid waters of the swamp.

Overcoming their superstitious fears, the Seminoles pushed onward until they came to the point where only a thick wall of tropical vegetation separated them from the island which was their objective, and with a savage yell and straining paddles they parted the foliage and dashed forward—

Only to halt in awed surprise and wonder.

For where had once been the fabled and beautiful island with its entrancing "Sun Daughters" was now an open space of swamp water, upon the surface of which a startled flock of wild ducks flapped and splashed before they took the air.

And from that day to this, the legend goes, no trace has ever been found of the "Land of The Daughters of The Sun," and no reason for its mysterious disappearance has ever come to light.

This is the quaint legend of the "Sun Daughters" and the disappointed Seminole lovers. There are many such, all spiced with the same flavor of mysticism and all as absorbing as an Arabian Night's tale.

There is the old story that the Okefenokee was originally inhabited by a race of giant aborigines seven and eight feet tall, whose mighty chieftain towered head and shoulder above his tallest warriors. Hordes of invading Seminoles killed out the giant race, according to the story, and took the swamp lands for their own.

Now one who is gifted in analytical reasoning could spend a pleasant evening pondering over the possible origin and meaning of the two legends. The "Sun Daughter" story would certainly suggest fire-worship or at least a kindred cult to his mind, and the legend of the giant people would probably lead to thoughts of a prehistoric race as ancient as the western cliff-dwellers.

Being of no scientific turn, I merely repeat the stories as they were related to me, and pass on to other items.

-Loyd A. Wilhoit

Commentary:

The story of the Okefenokee Daughters of the Sun has received some attention in cryptozoological and Fortean circles because it is claimed that their husbands were giant, hairy wild men and that the hunters could hear half-human, half-animal shrieks when they attempted to rediscover the Enchanted Island. These details are a much later addition to the original story. Neither William Bartram nor any other writer of the late 18th and early 19th centuries mentioned these aspects of the husbands. They simply describe the husbands as fierce, cruel, and aggressive. Another later addition to the story is that the island had castles that appeared Spanish in architecture, which has caused some to speculate that the Daughters were Timucuans who were in alliance with the Spanish. Again, Bartram does not include this detail, so it must have come from a later source.

In contrast to those later theories, Bartram uses another story told to him—though he does not elucidate what that story was—to suggest that these Daughters of the Sun and their husbands were a band of Yamasees who had found refuge in the swamp. This is the most plausible theory both because of what we know of Bartram and what we know of the Yamasees.

William Bartram was a botanist, an ornithologist, and the son of famous botanist John Bartram. He was primarily interested in studying the flora and fauna of Georgia and the surrounding states. Stories like the Daughters of the Sun were added as cultural commentary but were not his focus of study. Therefore, he had little motivation to lie or embellish. He was also a Quaker, so equality was one of his core tenets. This allowed him to interact with Native Americans in a more friendly manner than most of his compatriots. He would have had cultural knowledge beyond what was included in his text, allowing him to reach a reasonably accurate conclusion.

Additionally, the Yamasees were brutal. They often raided other Native tribes and sold their prisoners of war to colonists as slaves. They also spoke a language within the Muskogean family, so they would have been mutually intelligible with Creeks and Seminoles living nearby. The Timucuans, on the other hand, spoke a language unrelated to other languages in the region, so it would have been difficult for Timucuan women to communicate with men from a different tribe.

A Yamasee population in the Okefenokee is the most reasonable explanation for the origin of this legend, and there is no indication from Bartram that literal monsters were afoot in the Okefenokee.

The Murderous Monster

January 20, 1829 - Charleston Daily Courier (Charleston, SC)

[From the Milledgeville Statesman.]

A Giant Story.—There is a tradition among the Creek Indians that there is, in the trackless gloom of the Okefenokee Swamp, an Island of enchanting beauty, more blissful than any spot on earth. While it is generally thought this murky fen—this black sea of Avernus, contains nothing higher in the order of beings than countless armies of mosquitoes, snakes, frogs, and alligators, the Indians say that in the terrestrial paradise on this Island, there dwells a race of mortals of super human dimensions and incomparable beauty. This Island, tho' sometimes seen, is represented as inaccessible from the attribute which it possesses of locomotion; thus eluding approach—or from the ever varying labyrinths of fens and bogs by which it is entrenched and in which the bold invader is confounded who ventures too near this enchanted spot. Thus lost, in extricable sloughs, a few intrepid hunters were once saved from perishing by a company of women from this Island of surprising form and beauty, whom they denominate the *Daughters of the Sun*, or children of the great Spirit. Having kindly supplied them with refreshments, and pointed out to them a way of retreat, they admonished them to fly for safety—for that their husbands were fierce men, and cruel to strangers.

This legend we have hitherto regarded as fabulous; but Mr. John Ostcan, residing on the borders of this swamp, in Ware county, and some of his neighbors over the line in Florida, have become satisfied from ocular reality, and they so aver, that it is, mainly a *matter of fact!* We have this statement in writing, tested by a respectable witness, who has put the paper in our hand, containing the following facts—We beg the gentleman's pardon—truths, we should say.

Not long ago, two men and a boy, in the vicinity of this swamp, like our friend Paul Pry, "had a curiosity to know, you know," what could be seen by a two or three weeks pilgrimage into the accessible regions of this dismal empire. The season being unusually dry, they pushed their explorations far into the interior, and at the end of little more than two weeks, found their progress suddenly arrested at the appearance of the print of a foot-step so unearthly in its dimensions, so ominous of power, and terrible in form, that they were at once reminded of the legend we have mentioned above, and began seriously to apprehend its solemn reality. The length of the foot was eighteen, and the breadth, nine inches. The monster, from every appearance, must have moved forward in an easy or hesitating gait, his stride, from heel to toe, being but a trifle over six feet! Our adventurers had seen enough! and began to think of securing a retreat, without waiting to salute his majesty, not doubting but the other part of the story might also prove true—of his fierceness and cruelty. They happily effected their escape, returned home, and related the history of their adventures, and what they had seen of the "Man mountain." A company of Florida hunters, half horse and half alligator—nine in number, determined, a few months since, to make this gentleman a visit—to ascertain if he had a family, and his manner of living. Following, for some days, the direction of their guide, they came at length upon the track first discovered; some vestiges of which were still remaining, pursuing these traces several days longer, they came to a halt on a little eminence, and determined to pitch their camp and refresh themselves for the day. The report of their rifles, as one or two of them were simultaneously discharged at an advancing and ferocious wild beast, made the still solitudes of these dismal lakes reverberate with deafening roar. Echo, beyond echo, took up and prolonged the sound, which seemed to die away and revive in successive peals, for several minutes. The report had reached and startled from his lair, the genius of the swamp, and the next minute he was full in their view, advancing upon them with a terrible look and a ferocious mien. Our little band, instinctively, gathering close in a body and presented their rifles. The huge being, nothing daunted, bounded upon his victims, and in the same instant received the contents of 7 rifles. But he did not fall alone; nor until he had glutted his wrath with the death of five of them, which he effected by wringing off the head from the body. Writhing and exhausted, at length, he fell, with his hapless prey beneath his grasp. The surviving four had opportunity to examine the dreadful being as he lay extended on the earth, sometimes wallowing and roaring.

His length was *thirteen feet*, and his breadth and volume of just proportions.—Fearing, lest the report of their rifles and stentorian yells of the expiring giant should bring suddenly upon them, the avengers of his blood, they betook themselves to flight, having first secured the rifles of their headless comrades, and returned home with this account of their adventures.

The story of the report, as related above, is a *matter of fact*, and the truth of it is accredited, we are told, by persons living on the borders of this swamp, and in the neighborhood of the surviving adventurers.

Commentary:

This is a sensationalized story, so it isn't easy to know precisely what these people experienced. That said, it seems unlikely that they would invent such a creature in 1829—well before the cultural zeitgeist of the wild man and the Abominable Snowman had developed—so we may be able to tease some facts out of the exaggerations.

The footprints were indicated as being 18" long, 9" wide, and "terrible in form." While the size is larger than is typically seen in Georgia and is most likely a victim of sensationalism, the "terrible in form" description is interesting, as it may suggest a track with a different anatomy than a human or closely related hominid. Perhaps the track even had a divergent big toe like the footprints of a Skunk-Ape-type relict hominoid. In 1829, the residents of Georgia would not have been familiar with the foot anatomy of a chimpanzee or gorilla, so they would not have had the language to describe the foot shape of a pongid and may have considered it a deformity.

The description of rifle echoes that became softer and louder again also begs for an alternative explanation. There is little in the swamp to create an echo. The ground is very flat, and the trees and other vegetation absorb the sound waves that the rock of a mountain or cave would reflect. Plus, the nature of an echo precludes it from getting louder again once it has lost energy and amplitude. The hunters were not hearing an echo. Instead, they witnessed the mimicking sound of wood knocks from relict hominoids in the area.

As far as killing five men by wringing their heads from their necks, it's hard to say if there is any veracity to the claim. We must note, however, that relict hominoids, particularly of the Pacific Northwest Sasquatch type, appear to use a twisting motion to break tree branches. Some researchers have even suggested that Pacific Northwest Sasquatches use the same twisting motion to break the necks of prey animals. Thus, it would be reasonable for such an incident to occur, but it does seem out of character. Such aggression toward humans is extremely rare in the corpus of witness accounts.

Chapter Five: The Saga of Walker County

Home of John Limber and His Wild Man

Walker County in Northwest Georgia has the most historical reports of hairy hominoids in the state. Due to a confluence of factors, the reports span decades, outnumbering those of any other area; thus, Walker County has been given its own section of the book.

Walker County was founded in 1833 as part of the Indian Removal process from lands formerly belonging to the Cherokee. Between 1837 and 1854, the neighboring counties of Dade, Chattooga, Whitfield, and Catoosa were formed using parts of the original Walker County.

While Walker remained completely rural for many decades, it has recently become quite suburbanized as part of the Chattanooga, TN, metropolitan area. Today, Walker County has over 65,000 residents, and Hamilton County, TN, of which Chattanooga is the county seat, has around 365,000 residents. In the 1880s and 1890s, when the bulk of the activity documented in this book occurred, Chattanooga saw a population boom due to Reconstruction industrialization. Hamilton County, TN, increased from 23,000 residents in 1880 to almost 62,000 in 1900. During this same time frame, Walker County had less than 15,000 residents. This means every man, woman, and child could have had 20 acres to roam and never see another human.

Despite the increase in population over the years, Walker County remains rural and undeveloped in many areas because of the rugged terrain. Walker County sits on the boundary between the Cumberland Plateau and Ridge and Valley geographical regions. Lookout Mountain, a ridge within the Cumberland Plateau, stretches across the county, with valleys carved out below.

Due to significant limestone deposits, caves and caverns dot the region. Walker County has 149 caves, second only to neighboring Dade County's 164 for the most caves in Georgia. Some of the more famous caves in Walker are Ellison's, Pettyjohn's, and Frick's. Ellison's Cave is the 12th deepest and 52nd longest cave in the United States and has the two deepest cave drops (586 feet and 440 feet) in the continental United States. Pettyjohn's Cave is the 3rd largest cave in Georgia and is popular with modern spelunkers. Frick's Cave is home to a population of endangered gray bats and has Georgia's only population of Tennessee cave salamanders.

With so many hills, hollows, and caves in the county, there are numerous remote areas for relict hominoids to thrive. Still, hominoid reports only happen if people are willing to report it. While many newspapers across Georgia were dismissive of and even mocked wild man sightings, the *Walker County Messenger* was willing to report on the sightings with minimal use of the wild man as a political trope.

The 1884 Encounter

Period: Spring of 1884

Location: Walker County, Northwest Georgia

February 7, 1884 - Walker County Messenger (LaFayette, Walker County, GA)

It is reported by a man who lives on the spurs of Lookout mountain, that there is a wild man roaming about, who is of giant size and as hairy as a Newfoundland dog, as well as he can guess, about nine feet high, and will weigh in the neighborhood of 500 pounds—has eyes giving light equal to the moon—an appearance of the most frightful nature and growls equal to the lion causing people in that section to remain at home of nights with closed doors and well fastened. I guess the gentleman has been unchained, and we sinners who have big defiance to good commandments, had better look out. No man 'possum hunting while he roams the forest. Should the old fellow appear over this way soon there will be a change of schedule in regard to future anticipations.

—JOHN LIMBER

March 8, 1884 - Haralson Banner (Buchanan, Haralson County, GA)

Mr. Editor, do you reckon the tale about the nine feet, 500 pounds and hairy man with such big eyes and voice really true? If so, please locate the place you call Pond Springs. It may be close to me; and if it is, I shall hide myself in the cliffs and rocks of this mountain. Hoping you and your paper abundant success. I am

Yours Truly,

ASHER.

May 8, 1884 - Walker County Messenger (LaFayette, Walker County, GA)

Now John Limber, I don't want you to have anything more to do with that wild man. But lay aside your grip sack, pistols and chains and come out flat-footed and tell us what you think about free whiskey, free tobacco (for I know you love both) and I will have you appointed overseer next year.

—W. E. M.

June 5, 1884 - Walker County Messenger (LaFayette, Walker County, GA)

John Limber sent us nothing this week. Hal was in town Tuesday and said John was out hunting the wild man.

Commentary:

This encounter in 1884 was the world's introduction to Walker County. The story became national news and was featured in newspapers from several states. With its large size, glowing eyes, and loud vocalizations, this relict hominoid is a clear example of an Eastern Bigfoot type.

The First 1889 Encounter

Period: Spring of 1889

Location: Walker County, Northwest Georgia

January 31, 1889 - Walker County Messenger (LaFayette, Walker County, GA)

EDITOR MESSENGER:—The wild man has again made his appearance, this time nearer than usual. He was seen last week near High Point not far from Mr. Parrishes store, by a reliable gentleman of undoubted veracity, who was searching along the spurs of Lookout Mountain for mineral indications. He had discovered several places that satisfied him of rich veins of iron ore and stone coal, which satisfied him that further search was unnecessary; therefore he turned his tramps in the direction of the valley. On arriving at the head of the hollow just above Mr. Olivers, his attention was attracted by very thrilling screams on the side of the high spur that he had just passed along the trail below. He stopped and looked in the direction from which the noise came. What should he see but the wild man that has been seen before, but not in that neighborhood, and never so close to any body's house. For a moment the hair on his head stood straight up and it was impossible for him to keep his hat on, or decide whether he was living on earth or had fled to the spirit land. When his fright had cooled down a little, he discovered that the old fellow was not making any move towards him, but was standing erect and shaking his fist at him and showing signs that were not favorable toward a human being. Therefore get way was thought about immediately, but before starting he thought he would examine the wild gentleman more minutely. He was about one hundred yards from him, but could distinctly see his teeth, nose and eyes, and describes him thus: Was about 7 or 7 1/2 feet high, hairy as an old bear and would weigh from his looks, four hundred pounds; had a pole in one hand that looked to be ten feet long which he handled as easily as a stout healthy man would a pipe stem. His name was asked and the answer came in the shape of a large stone which weighed at least one hundred pounds, which was hurled at the inquisitive gentleman with the force of a cannon ball. This being done, there was no time left for any further inquiry, consequently he tried the working speed of his legs and feet which worked most excellent until he arrived at Frank Carter's shop, almost breathless. As soon as he could narrate his story, a crowd was collected which started in search of the wild man of Lookout Mountain.

They soon arrived at the spot where the gentleman saw him, but he was gone only leaving signs of where he had been by breaking down saplings, turning over large rocks and tearing up the earth. Not being eager for farther search, they returned to their respective homes, leaving the wild man of Lookout to roam at will.

—JOHN LIMBER

February 11, 1889 - Walker County Messenger (LaFayette, Walker County, GA)

CHATTANOOGA, Tenn., Feb. 7.—The citizens of Walker county, Georgia, a few miles from this city, are very much excited over the existence of a genuine wild man, who haunts the mountains of the county. He is described as being of gigantic stature, covered with a thick growth of hair, and he carries in his hand a large knotted stick. He looked as if he might be the twin brother of Barnum's wild man.

This modern Orson has been seen by several parties. One gentleman bolder than the rest encountered the creature in a lonely part of the mountains one day not long since and at a safe distance endeavored to strike up a conversation. A perfect shower of rocks greeted his first words, and thinking discretion the better part of valor, he made tracks from the dangerous neighborhood.

February 14, 1889 - Walker County Messenger (LaFayette, Walker County, GA)

Let John Limber keep the wild man on Lookout. We don't want him on Taylor's Ridge, especially on our part of it.

—HENRY

March 7, 1889 - Walker County Messenger (LaFayette, Walker County, GA)

I was informed yesterday, that Harve Hollingsworth is making arrangements to capture the wild man, as follows: He has seen him at one place at different times recently, and he is now building a wire fence around the spot where the old chap seems to be located. Well, he is certain, for dirt is banked up three or perhaps five feet high, which came from the hole he scratched in the ground. As soon as the fence is completed, he will couple gas pipes together sufficiently long to reach the spot of the desired object. This done, he will with a force pump, force three or four barrels of chloroform into that hole through the pipe, which will cause a stupor to come on the old fellow. Then with heavy wire ropes he can be tied. However all precautions will be taken to keep at a safe distance after the tying is done, as on awaking he might tear things up generally. The wire fence is only to tangle him that distance might be gained in case of a run. See Eb and borrow his gas engine to force your chloroform through the pipe, as he will not need it till he completes his smelter.

If you see Frank Carter in your rounds, say to him that Uncle John finished up the last pair of tongs Saturday and is now ready for business. Now, Harve, be careful or that hairy man will get you and your assistants that are aiding you in the venture. Hope to hear of your success.

—JOHN LIMBER

March 14, 1889 - Walker County Messenger (LaFayette, Walker County, GA)

John Allison and old man Oliver have engage a large amount of hands to fill up the holes that John Limber's wild man tore out in the side of Lookout Mountain.

—SAMBO

March 14, 1889 - Walker County Messenger (LaFayette, Walker County, GA)

I was in the other Valley a day or two last week, found everybody peaceable and quiet, making extensive preparation for farming it right this season. They all take the MESSENGER, which accounts for their happiness and prosperity. The wild man has given them the slip, consequently the excitement in regard to his whereabouts is over. Eb found the hammer than he used in knocking the loose rocks out of his way on the mountain, which can be seen any time at his mill, lying in close proximity to that wind machine he had made recently.

Some unknown creature of not much respect for the welfare of others, entered Bud Harris' house last week one day in the absence of the family—took some jelly etc., tore a quilt and sheet in pieces and somewhat injured the bed tick. No one saw the animal, though there were some hands plowing near by. The supposition is that the wild man has left the other valley and located in the cedar thicket near by here, as rumblings are frequently heard after dark, near the pump. He escaped from Harve and he does not know which way he went.

—JOHN LIMBER

March 21, 1889 - Walker County Messenger (LaFayette, Walker County, GA)

I have noticed John Limber has been writing something about a wild man on old Lookout Mountain and in his last letter he said he thought Harvey Hollingsworth would likely capture him. I hope he will be successful, and I want to hear all about his being captured and how Harvey did it. I guess Frank will assist Harvey. I hope they will succeed without getting a scratch or losing any blood.

—M. D. ALLISON

MARCH 28, 1889 - Walker County Messenger (LaFayette, Walker County, GA)

If they will let us know when they go to capture the wild man, if they should need any help we will go up and assist them.

—MARION

April 4, 1889 - Walker County Messenger (LaFayette, Walker County, GA)

The wild man was captured the other day, and is now in close confinement; however he is quieting down some. He refuses to eat anything but wild meats, such as rabbits, coons and frogs, that too without any cooking or cleaning are thrown into his cage just as caught and are devoured immediately.

—JOHN LIMBER

July 11, 1889 - Walker County Messenger (LaFayette, Walker County, GA)

Ed Howard tell me that everything in the other valley is moving along very nicely—crops are good—peace and happiness reign supreme, save the finding the wild man's track occasionally which creates a little excitement.

—JOHN LIMBER

Commentary:

As seen in this and the next two encounters, 1889 was by far the most fruitful year for wild man sightings in Walker County. John Limber thrived as a reporter of these tales, and "John Limber's wild man" became the common term for relict hominoids in the pages of the *Walker County Messenger*.

The immense size, superhuman strength, and hairy appearance again make this account a clear example of an Eastern-Bigfoot-type relict hominoid. This conclusion is further validated by the wild man's rock-throwing behavior. Such conduct is notoriously reported in the Pacific Northwest Sasquatch type but is also commonly witnessed in their close relative, the Eastern Bigfoot.

The Second 1889 Encounter

Period: March of 1889

Location: Walker County, Northwest Georgia

March 9, 1889 - The Tennessean (Nashville, TN)

CHATTANOOGA, March 8.—Walker County, Georgia, in the vicinity of Chattanooga and as far south as Pond Springs, is all torn up over the reappearance of the celebrated wild man of Lookout. He was seen a few days since, and, if descriptions are correct, he is a most remarkable being. His hair and beard are described as flowing to the waist, his finger and toe-nails are long, giving the hands and feet the resemblance of claws. He wears a trunk of bear skin, with a bear-skin robe thrown over his shoulders. He carries an ugly bludgeon and persistently avoids coming in contact with anybody. The timid people of the neighborhood are greatly alarmed and there is little traveling about at night, although he is generally believed to be harmless. This strange creature has been haunting the caves and fastnesses of Lookout Mountain and elevations in lower East Tennessee for years, and nothing is known of his identity. He has never been known to do anybody any harm and there has been no occasion for his arrest, so he has been allowed to pursue his strange lunacy unarrested. He is said now to occupy a cave near Pond Springs, Ga., but heretofore when his place of temporary abode has been discovered he has disappeared, to be seen at other point a considerable distance away.

Commentary:

Given that this account is from a newspaper so far from Walker County, it would be easy to dismiss it as a corruption of the earlier sightings of the Eastern Bigfoot. However, it is such an accurate description of a Neanderthaloid type that it should not be dismissed out of hand. The long hair, long beard, claw-like nails, and animal skin clothing are all quintessentially Neanderthaloid.

The Third 1889 Encounter

Period: October of 1889

Location: Walker County, Northwest Georgia

October 24, 1889 - Walker County Messenger (LaFayette, Walker County, GA)

While at my old friend Eb's in the other Valley, last week, I learned that Elie Woodall was on the mountain side Sunday previous, and found the wild man about whom a previous account has been given in the columns of the Messenger. He discovered him in the trunk of a very large chestnut tree, shaking chestnuts therefrom into his old lady's apron who was standing with it outstretched to receive them as they came rattling down. Elie being somewhat of a timid nature, did not venture close enough to let himself be known, but turned himself about and lit out down the mountain for his home, taking all near cuts regardless of rocks, brush etc., and now is troubled with pains and aches, also his pants are rather dilapidated, the brush having disorganized their seams. Elie, be of good cheer—report to Andy and obtain relief, though he be in much trouble about the loss of his Wilson turbine by the recent high waters, he will aid a poor distressed fellowman, as Eb proposes to share him in the loss.

—JOHN LIMBER

Commentary:

John Limber fails to give us any physical description for this supposed wild man. We are left with only behavior to determine what Elie Woodall saw. If Elie Woodall indeed saw a relict hominoid, the fact that the female was wearing and using an apron suggests the pair were Neanderthaloids. As Walker County had both Eastern Bigfoot and Neanderthaloid encounters in 1889, more information is necessary to determine whether it was a relict hominoid and, if so, what type.

The 1890 Encounter

Period: September of 1890

Location: Walker County, Northwest Georgia

September 5, 1890 - Columbus Ledger-Enquirer (Columbus, Muscogee County, GA)

It is again reported that a wild man has been seen near Snodgrass Hill, on Chickamauga battlefield. John Morris, a farmer living near there, obtained the closest view of him. He says that he was passing through the field below Snodgrass Hill when he heard something running in the bushes. He looked and saw a man, stark naked with hair all over his body, finger and toe nails looking like the claws of an animal. When the strange being saw the farmer he arose and with a cry darted through the woods and was soon lost to sight. The only explanation as to the wild man of Chickamauga is that he is an escaped convict. There have been several escapes from below Crawfish Spring and one of them was caught last Friday while trying to break his shackles with a rock. Whether the wild man is actually wild or is playing the wild man until he succeeds in getting a suit of citizen's clothes is not known. He has been seen only by two or three persons, so far as known, and they have been too badly frightened to give a very intelligent description of him, except that he is a white man and has no clothes. The citizens will probably capture him and unravel the mystery.

Commentary:

This wild man is most likely a Neanderthaloid-type relict hominoid. The first indication is that he was assumed to be human despite being covered all over with hair. The second indication is that his nails were described as claws, a defining feature of the Neanderthaloids.

The 1891 Encounter

Period: May of 1891

Location: Walker County, Northwest Georgia

May 29, 1891 - Columbus Ledger-Enquirer (Columbus, Muscogee County, GA)

A party of negroes report seeing a wild man near LaFayette recently. He was in an almost nude condition, his hair almost hiding his face and reaching to his waist. On being approached he uttered a low growl, like that of a dog, and fled into the woods. The negroes were frightened, and also took to their heels.

Commentary:

While this account is short, it does seem to describe a Neanderthaloid-type relict hominoid. The long hair to the waist is the most striking feature suggesting a Neanderthaloid origin, though it should be acknowledged that a mere human cannot be excluded based on the limited details.

The 1894 Encounter

Period: Spring of 1894

Location: Walker County, Northwest Georgia

April 26, 1894 - Walker County Messenger (LaFayette, Walker County, GA)

My old and esteemed friend, Eb Carlock in the other valley, has been very much frightened for the past few days, caused by a rumbling on the mountain side beyond his house. It may be a volcanic eruption, or perhaps that wild man has made his appearance again on the mountain—is pulling up trees and rolling huge stones from their present location down the mountain. I would suggest that Eb call Uncle Jo and Bud in for consultation the first rainy day, and formulate some plan by which the cause of all this rumbling originates. Procure a field glass—get on the highest spur of the mountain and look carefully around and be sure it is not the wild man before getting too close. If the cause be ascertained, it should be given to the public through the columns of the Messenger.

—JOHN LIMBER

May 3, 1894 - Walker County Messenger (LaFayette, Walker County, GA)

We have not heard or seen anything of the wild man this week, but Coxey's army is on a tare and the world is retrograding, so say the Cooper Height boys.

June 7, 1894 - Walker County Messenger (LaFayette, Walker County, GA)

The cold weather has damaged cotton and gardens. It has also hurt John Limber's wild man. The last I heard of him he was still roving around muttering and growling. Bud and Ebb are fixing some way to tame him. There is no danger in him.

Commentary:

Despite John Limber's suggestion, no volcanoes exist in Georgia or the surrounding states. There were also no earthquakes recorded in the Southeast in 1894. We have no direct evidence that relict hominoids caused whatever "rumblings" Eb Carlock and his neighbors experienced; however, the possible explanation of pulling up trees mirrors the actions of the legendary Tsul'kălû' of Cherokee lore. If a relict hominoid was responsible for such loud shenanigans, it would almost assuredly be an Eastern Bigfoot type.

The 1900 Encounter

Period: July of 1900

Location: Walker County, Northwest Georgia

July 5, 1900 - Walker County Messenger (LaFayette, Walker County, GA)

I noticed in the Times that the Lookout Mountain wild man had been seen again. I saw Eb last week who lives in the neighborhood of where he was seen, but he failed to tell me anything about the discovery. If Andy could get close enough to him to touch him with a small amount of his salve, he could be easily caught and tamed.

—John Limber

Commentary:

Sadly, the initial report from the *Chattanooga Times* could not be located. Based on previous reports from Walker County and Tennessee newspapers, both Eastern-Bigfoot-type and Neanderthaloid-type relict hominoids could be found in Eb Carlock's area of the county. Without further description, we cannot differentiate.

The 1903 Encounter

Period: February of 1903

Location: Walker County, Northwest Georgia

February 19, 1903 - Walker County Messenger (LaFayette, Walker County, GA)

John Limber's wild man is now located on J. B. Faucett's place.

Commentary:

We can suppose that this might be another relict hominoid sighting as it is described as John Limber's wild man, recollecting back to the work John Limber did in reporting the wild man sightings of the 1880s and 1890s. Unfortunately, by this time, "John Limber's wild man" was also used in jest by the paper, so without additional information, there is no context to judge this report.

The 1904 Encounter

Period: March of 1904

Location: Walker County, Northwest Georgia

March 18, 1904 - Dade County Sentinel (Trenton, Dade County, GA)

From Sitton's Gulf comes a strange and thrilling story of a recent discovery made there. The report is so startling and uncommon that by many it is entirely discredited. Some hunters, while enjoying a round in the wilds of this well known depression of Lookout mountain, came unexpectedly upon signs of human habitation. From a yawning cave near by, to whose entrance led several beaten paths showing human footprints in the loose surface dust, came the sound of a human voice. The hunters, be it said to their bravery and composure, hastily retreated down the slopes sufficiently to find a location for a stand, while carrying their guns in one hand and reaching for their suddenly ascending hats with the other. No sooner had the gallant boys halted and faced the now terrible expostulations and threats emanating from the caverns, than there appeared a form of such weird appearance, at the mouth of the cave, the blood of the hunters was chilled. Time and time again did they attempt to raise their guns to shoot but, lo and behold! the weapons dropped from their hands. "It's a wi-ild m-man!" cries one, while the inhabitant of the cave jumped towards them, and started a huge bolder down in their faces. Vacation then suddenly became the marching order, and the hunters and the large bolder vied with each other in a terrific race for more desirable quarters. Guns, hats and wearing paraphernalia were left in confusion to the keeping of the wild man of the Gulf. A searching and constructing party will be sent out to find and exterminate the intruder, and prepare quarters for all defeated candidates, in the near future.

Commentary:

This wild man is given no physical description. Despite the lack of information, the individual was almost assuredly an Eastern Bigfoot type because of the distinct rock-throwing behavior. Such behavior is often seen in encounters with the Eastern Bigfoot and the Pacific Northwest Sasquatch types but is not reported in encounters with other kinds of relict hominoids.

The 1906 Encounter

Period: November of 1906

Location: Walker County, Northwest Georgia

November 9, 1906 - Walker County Messenger (LaFayette, Walker County, GA)

Say, has the wild man been caught that has been roving around Ascalon so long?

—A.F. SHAW

Commentary:

The records of the *Walker County Messenger* did not include any other references to the wild man in Ascalon (an unincorporated community of Walker County) in 1906, so this may have been a local rumor that did not make its way into the papers.

The 1907 Encounter

Period: July of 1907

Location: Walker County, Northwest Georgia

July 25, 1907 - Walker County Messenger (LaFayette, Walker County, GA)

Last week some negroes of this place while picking blackberries near Lytle discoveries a wild man hiding in a blackberry patch. He was followed about two miles south and all trace of him was lost near the Dan Shields' place. He was rough-dressed, had dark red hair and beard and looked to be about 40 years of age.

Commentary:

This is a relatively vague account that is by no means a definitive example of a relict hominoid. It is, however, Walker County, so anything is possible. The dark red hue of both hair and beard, as well as the ragged appearance of his clothing, is consistent with the profile of a Neanderthaloid.

This marks the end of the Walker County saga. While sightings of hairy hominoids continued, they were no longer reported by the *Messenger* as a wild man sighting.

Chapter Six: Other Historical Accounts

94

The Yahoo of the South Carolina Border

Period: Unknown

Location: Habersham County, Northeast Georgia and Oconee County, South Carolina

June 5, 1889 - **Savannah Morning News (Savannah, Chatham County, GA)**

A correspondent of the Clarkesville Advertiser relates the following incident connected with the early history of Habersham county: "During the time the Indians were in the south a hunting party established a camp a little east of Tugalo river, in what is now Oconee county, South Carolina. One day they all went hunting, leaving a deer they had killed the evening previous, at the camp. At night when the Indians returned to the camp the deer was gone, and the next day the same thing was repeated, when they concluded to leave an old Indian to guard the camp and see what went with their deer. That day the old Indian saw a monster animal come and carry off the deer, and was afraid to make any attempt to kill the monster, which was about 7 feet high and walked erect like a man, hairy all over, and its mouth was in the chin, and great claws on the fingers and toes. The next day all seven of the Indians stayed at the camp and, as usual, the monster came, gathered up the deer and started off, when one of them fired at it, the ball taking effect in the back. The animal dropped the deer and turned and started toward them, when the other six fired a volley into its breast and it fell dead. About three hours after that the Indians heard a noise like some one hallooing about a mile distant. 'Yaho, yaho, yaho!' The Indians left the camp and called on the *posse comitatus* for protection, when a party of whites on horses, with all the dogs they could get, went in search of the other animal and found it. It was like the one the Indians had killed, and putting the dogs after it, when the men appeared in sight the animal would run, but it could whip every dog they could get after it. The party pursued it to the river, and at two jumps it went across the river over into Habersham county, and was shot by a party soon after it crossed the river."

Commentary:

There is no date given for this account. Still, because of the prevalence of Native Americans in the narrative, we can assume that it probably occurred before the Indian Removal era, which began in 1815. Habersham County, Georgia, was not founded until 1817, and Oconee County, South Carolina, was not founded until 1868, so there are likely no contemporary records of *posse comitatus* activities related to this account.

Given that the rough description of the first hominoid is that of a 7-foot-tall, hairy, man-like creature, it seems evident that we are dealing with an Eastern Bigfoot type. The second hominoid is described as being "like the one the Indians had killed," so it likewise would be an Eastern Bigfoot type. For size comparisons, it would be helpful to know how wide the river was where it crossed in two leaps. Unfortunately, that area is now underwater. Most of the Georgia-South Carolina border north of Augusta is today a series of artificial lakes created by hydroelectric dams built in the early 20th century.

This account has several exciting features, including that two hairy hominoids were killed, the "mouth was in the chin," the first hominoid was stealing deer from camp, and the call sounded like "yaho."

Skeptics might chalk up a sighting of just one hairy hominoid to an overactive imagination or a mistaken identity. That type of criticism becomes more tenuous when there are two hairy hominoids. It becomes even more tenuous when these creatures are killed, meaning there is time to examine them closely after death. Given the circumstances, the probability of misidentifying a bear or other known animal is minimal.

The description of the hominoid's mouth being "in the chin" might seem nightmarish, but it describes a jaw structure similar to that of a gorilla or the Pacific Northwest Sasquatch. Dr. Jeff Meldrum describes the jaw structure of Patty from the Patterson-Gimlin film as an "extremely robust chewing apparatus." The larger jaw muscles and high attachment points necessary for grinding through vast amounts of vegetation daily create the appearance of a hominoid with little to no neck and a mouth that sits lower on the face.

The jaw anatomy is one reason some researchers believe the Pacific Northwest Sasquatch type is vegetarian. The Albert Ostman kidnapping account seemingly verifies that hypothesis, though Ostman is not without his skeptics. In Georgia, however, this account of Eastern Bigfoots stealing game suggests that the Eastern Bigfoot type lives off a diet similar to that of the hypocarnivorous black bear, where they primarily eat vegetation but opportunistically feed on meat and other protein as the situation presents.

The "yaho" call is typical of relict hominoids, to the extent that Yahoo became a nickname for the Eastern Bigfoot type. The Cherokee legend of Yawâ'ĭ discussed in a previous chapter corroborates this.

The Blue Monster of Rabun County

Period: August of 1812

Location: Rabun County, Northeast Georgia

May 3, 1891 - Atlanta Constitution (Atlanta, Fulton County, GA)

In the fore part of August, 1812, a party of hunters found in a mountainous region now known as Rabun county, Ga., a being nearly eight feet high covered with bluish hair and having a human face adorned with immense ears resembling those of an ass. The creature was stone deaf and on that account seemed wholly unconscious of the approach of the men. This monster seems, from old accounts, to have been seen upon several occasions during the next four years.

In 1816 a number of adventurers from Virginia, most of them surveyors working up the unexplored portions of Georgia and the Carolinas, formed themselves into a party for the express purpose of capturing the uncanny being if possible. They scoured the hills and valleys for several days and at last returned unsuccessful to the starting point. The learned Joseph Earle, then living at Culpeper, Va., wrote the following in a letter to John Bishop, of Boston, Lincolnshire, England: "An awful creature, half animal and half man, of gigantic stature and fierce mien, is known to inhabit the wild regions to the south of us. Some think that there is a race of these monsters hiding in the hills and mountains of Georgia, the place where it or they have been the oftenest seen. Of the few people which inhabit this wild country, not a soul which we have approached doubts that the creature is all that it is represented as being. Indeed, sir, one poor planter, who guided us a great distance from the falls, is convinced that he saw him face to face not more than three weeks since, an assertion my adventurous companions were only too ready to believe." The many tales of this extraordinary being seem to have created quite a stir all along the Atlantic coast. A printed circular issued by a land company in 1815 says, "The climate of Georgia is exceedingly mild, the soil productive, and the danger of attack from uncouth beasts which are represented as being half beast and half man are fairy tales not worthy of consideration."

Commentary:

Based on the large size of almost 8 feet, this creature falls into the Eastern Bigfoot category. However, a couple of characteristics defy the typical Eastern Bigfoot profile.

The bluish hair described in the account is highly unusual for any hominoid. It is most easily explained by a deep black color that appears bluish in the sun due to blue undertones, similar to the color of a raven's feathers. While a deep black coloration is less common in the Eastern Bigfoot type than a brown or dark brown hue, it has been reported.

The ears like an ass and associated deafness are more challenging to explain. Some ear deformities in humans are associated with hearing loss, but they do not include a donkey-like ear shape. The most likely cause of ears that appear pointy like those of a donkey is a condition known as Stahl's ear, where an extra fold of cartilage creates a pointed appearance. Stahl's ear, however, is not associated with hearing loss. It's possible that the deafness was separate from the ear shape, but perhaps it was not a medical condition. Animals that do not have much exposure to humans will often allow humans to get very close to them, as they have no fear. Perhaps the creature simply did not fear the humans that approached him.

If one day, a copy of the land company circular or the letter from Joseph Earle to John Bishop is recovered, this account will gain substantial credibility. That day, however, will likely never come. Rabun County was not founded until 1819, and the local library and historical society have no records from before that time. For now, it is hearsay of an encounter documented 80 years after the fact.

The Mesmerizer of Fannin County

Period: May and June of 1873

Location: Fannin County, Northwest Georgia

May 23, 1873 - Cleveland Banner (Cleveland, TN)

A Monster at Large.

EDITOR OF THE BANNER—Dear Sir: The horrible sight of a hairy man has been seen in Fannin county, Georgia. His is wild and monstrous—he has been seen in houses carrying off women and children. He is eight feet high, and is covered all over with black curly hair. He started from a house lately with a woman in his arms, but by the approach of two men she was released. The settlement was alarmed and pursuit given on horseback. After a hard ride the monster was overtaken and a terrible fight ensued, in which a man by the name of John Haircrow was killed, and a horse had his tail torn off, and the pursuers were forced to retreat and leave the field in possession of the monster. The settlers are arming themselves with guns and watching for him.—He makes his appearance just before or in time of a rain.

Yours, truly,

DAVID A. BALLEW, JR.

Wolf Creek, Cherokee co., N. C., May 13th

June 4, 1873 - Savannah Morning News (Savannah, Chatham County, GA)

We copy the following from the Western press, where it is highly popular: The wild man in the woods has made his annual appearance once more, and very early, he being this time in Fannin county, Georgia. He is represented to be eight feet high, and in a recent fight with some of the citizens who attempted to capture him, he killed one of them, as represented, and tore off the tail of a horse.

June 12, 1873 - The Cambridge City Tribune (Cambridge City, IN)

The wild man in the woods is in Georgia now. He has got to be eight feet high, and tears out a horse's tail with one turn of the wrist.

June 13, 1873 - Cleveland Banner (Cleveland, TN)

The Monster Still at Large in Fannin County, Georgia.

ED. BANNER: Dear Sir: The horrible sight of the hairy man, given an account of in the Banner 23d ultimo, by D. A. Ballew, is creating a wild excitement in this county. And still more horrible to relate, there have been seen in company with him a Dwarf man and a large wild Bull. The Dwarf man is slightly inclined to be hairy, his eyes look like balls of fire, appears to be about 5 3/4 feet high, very lean, and supposed to weigh not more than 36 pounds. In breathing he makes a small whistling noise through his nose that can be heard a quarter of a mile. This strange breathing is so loud as to warn the excited women and children of his approach in time to hide. The Bull is supposed to weigh 2,000 lbs, is wild and furious and can run as fast as a fleet horse. The monster hairy man, as Ballew informed you, is eight feet high, and supposed to weight 400 lbs, and is equal in speed and strength to the Bull. The Dwarf and Bull seem to be under the supreme control of the hairy man. The hairy man and Bull have been seen on the high sunny points of the Crackers Neck Mountain licking each other.—They have also been seen to make several circuits around the summit of this mountain at a very high speed, making a noise like a storm of wind, seeming to be sporting.

The hairy man passed through the settlement and picked up a cast off jewel, which he seemed to prize very highly as he has often been seen fondling and amusing himself with it.

These monsters seem to possess supernatural powers; they can cause ghosts to be seen anywhere in the settlement; making the most horrible noises ever heard by man. A widow lady and two lovely daughters have been frightened by these monsters until they are almost maniacs. This widow and one of her daughters, a handsome young woman, were mesmerized and carried a half mile from home, in the dead hours of the night, they not being sensible of their removal, and shut up in an old dilapidated house, standing hear a dismal swamp, at the head of a mill pond. When they awoke from the mesmeric spell they heard a thundering crash come down from above; the roof of the old house was rung off, they heard the strange breathing of the Dwarf man high up in the elements; in a moment he came down through the house top, saying, "peace, dear widow and daughter, I am a wire-worker, and have power even to move regiments and am acting under orders of the hairy man." Remaining here three dreadful hours, listening to his incantations, frantic with fear, they were suddenly mesmerized and mysteriously carried to their home. After remaining at home, they knew not how long, the mesmeric spell was removed, and they heard the strange breathing of the Dwarf on the top of the chimney, repeating his incantations. On opening their gaze was the monster hairy man fondling his jewel. The daughter left at home was in a mesmeric spell during their entire absence, and knew nothing of the strange proceedings, excepting a faint recollection of hearing the monster, in a soliloquy, express fears of some other being, of higher power stealing away his precious jewel.

Such noises and sights as have been heard and seen about this swamp, have never before been heard in Georgia. In passing near a shop at a very high speed the hairy man ran against a two-inch auger, which inflicted a ghastly wound—he raged like an angry tiger—gnawed a lane through the bushes, and made the most unearthly howls ever heard by a human. When his rage somewhat abated, he directed his course where there were two invalids, a male and a female, and by his agonies and gestures, they understood that he wished his wound dressed, which they proceeded to do the best they could, in their excited condition. While dressing it, they remarked that the wound appeared to have been inflicted with a two-inch auger; as the word auger was articulated, he vanished like a meteor. Since this strange occurrence, these monsters flee at the sight of a two-inch auger.

The citizens are contemplating the construction of a trap, for the purpose of capturing these monsters without wounding them, hoping thereby to sell them to Barnum for his museum. Don't you think he would pay a handsome price for them? At least a sufficient sum to remunerate us for their capture and transportation to New York?

As further developments are made concerning these monsters I will write you.

A CITIZEN OF FANNIN CO., GA.

Commentary:

This account begins reasonably enough but soon ventures into the fantastical. Unfortunately, there is little corroborating evidence for even the initial report. David A. Ballew, Jr. of Cherokee County, NC, certainly existed, but John Haircrow, the man supposedly killed in the struggle with the wild man, is a mystery. No obituary or grave record seems to exist for Mr. Haircrow. His surname suggests Cherokee heritage, but there are several potential Johns in the rolls of the Eastern Band of Cherokees and no indication in the rolls of where each was living.

The follow-up report from the anonymous citizen of Fannin County is incredible, both in the sense of being extraordinary and difficult to believe. Adding the Dwarf Man and Bull to the narrative magnifies the confusion, as, for example, no living being of 5'9" should weigh just 36 pounds. 136 would be a far more biologically reasonable number, so perhaps a typo occurred, but even still, it is an oddly specific number. On top of that, the abduction tale from the two women makes the controversial Albert Ostman account of British Columbia in 1924 seem downright dull.

Realistically, this is probably just a sighting of an Eastern Bigfoot type with no abduction and possibly no killing of John Haircrow. If the Dwarf Man did exist, it was likely a juvenile Eastern Bigfoot.

"The Long-Haired Beast Eater of North Georgia"

Period: February of 1883

Location: Bartow, Catoosa, Dade, Walker, and Whitefield Counties, Northwest Georgia

February 1, 1883 - Chattanooga Daily Times (Chattanooga, TN)

For several weeks past various reports have reached the city of the presence of a wild man along the line of the Western & Atlantic railroad. It is said that two weeks since he was seen near Cartersville devouring the remains of a horse. He was perfectly nude, and before he left the carcass he skinned it and encircled his body with the skin. His beard was four or five feet long, and his unkempt hair hung far beneath his shoulders. He was very tall and strongly built, and had the air of a perfect demon. When next seen he was near Chicamauga, about eight miles from this city, devouring a dog's carcass. He carried a skillet with him, and when possible would cook his meat, but only on rare occasions.

Tuesday night he boarded a freight train on the road and succeeded in terrifying the crew, but they were enabled, by careful management, to secure him in a car and carried him to Cartersville, where he was lodged in jail. He wore nothing but the horse-skin, and on his little finger a gold ring, firmly imbedded in the flesh. He answers no questions, and his identity is a deep mystery.

February 15, 1883 - Oglethorpe Echo (Crawford, Oglethorpe County, GA)

Startling intelligence was received last night from Cartersville giving the details of the capture and incarceration of the demoniac yahoo whose baneful presence has been a source of alarm and terror to the inhabitants of the State living along the line of the W. & A. R. R. The ferocious savage after paralyzing the bravest bloods of Acworth on Monday as detailed in these columns, passed on unmolested towards the hyperborean North. He was next heard from at Chicamauga where he desecrated that far famed field of battle by devouring the putrid carcass of a dead dog that had been killed about a week preceding.

The inhabitants of that classic region, though a brave people, were not prepared to tackle the monster without a direct and peremptory order from the government, and the fiend in consequence ranged around the neighborhood at his own sweet will. After devouring the dead dog the yahoo turned his attention to a forest of fine persimmons, a fruit that finely flourishes in that section, but upon eating one he left the spot, apparently disgusted with their flavor, and proceeded to discuss the merits of a decomposed buzzard, whose feathered frame lay rotting by the road side. The historic bird, with its sable plumage, was devoured in ten minutes time, and the insatiate monster picked his teeth with the tail bone.

Having thus feasted himself, he squatted on his haunches, preparatory to refreshing himself with a sweet sleep. He remained in this position so long as to excite the astonishment of several persons who had been watching his manoeuvres from a distance, and they were emboldened by witnessing his harmless attitude to venture nearer. Two of the bravest of the number crept up noiselessly to within a hundred yards of the carrion savage, when they observed that his eyes were closed and that he was apparently in deep slumber. After satisfying themselves upon this point they proceeded to a critical examination of his person and effects.

The former they found to about tally with the description of him which they had read in the Post-Appeal. His beard, of a glittering, reddish hue, was about five feet two inches in length and hung in graceful neglect around his enormous flank, which looked to be about eleven inches in circumference. The volute curveture of the spinal cords of his legs were found to have been accurately described. His feet were some sixteen inches long, but a singular phenomenon was observed about his toes, the interstics of which actually contained a vegetable growth in appearance much resembling the rag weed of our Southern clime.

When his feet were placed close together, they formed a unique and original green plot, much resembling an Atlanta park, only a little larger. The soil that feeds this anomalous growth is well fertilized by the odiferous moisture emitted by those small members that form the extremity of the foot. His horse hide lay in repose beside him, and near it was observed an antiquated skillet, the contemplation of which latter suggested the harrowing possibility that the strange being might be a candidate for Governor in disguise.

"He evidently toteth his own skillet," mused one of the men, as he eyed the historic utensil meditatively. At the sound of his voice the Yahoo moved as if about to awake, and they beat a hasty retreat.

They returned to the village and organized an armed band of fifty men, with the intentions of capturing the human carrion, but when they reached the spot the savage had departed. On Wednesday night, about 9 o'clock, a freight train stopped near Chicamauga to water and wood up, which favorable opportunity was seized by the monster, who lurked in the vicinity, to board the train and get a free ride. The train moved off with the demon on board. The first man apprised of this fact was a brakeman, who was so paralyzed by the shock when he confronted the bivulpian horror that he fell from the moving train in a swoon, and has not been seen. The whole crew of the train within a half hour had learned of his presence, and after recovering from the paralytic shock that seized them when he was discovered, nine lariats, three carbines and a pistol were brought into requisition to effect his capture. He was carefully tied to a brake on the top of a box car and placed under a guard of two men armed with Winchester sixteen repeaters and orders to shoot him on the spot or in the head if he attempted to make his escape.

In this manner he was safely transported to Cartersville, where the authorities were informed of the hairy demon's capture, and requested to give him a lodgement in the county jail.

In an hour's time they succeeded in placing the wild man behind the bars, where he exhibited all the characteristics of a caged hyena. He was perfectly nude, wearing nothing but the historic horse hide. On his little finger a gold ring was seen deeply embedded in the flesh. Several questions were asked him but he declined to speak or explain his identity. A mysterious gloom surrounds his origin, which will probably be dissipated by the electric light of a legal inquiry by the properly constituted courts of the country.

February 3, 1883 - Chattanooga Daily Times (Chattanooga, TN)

News reached this city yesterday that the wild man who was captured at Cartersville a few days since has escaped.

Thursday evening a sympathizing gentleman volunteered to give the man a suit of clothes if some one would induce him to throw away his horse skin raiment. This was done, and he was accordingly taken to the Etowah river and, while a negro was bathing him, he knocked him down, and plunging into the river swam to the opposite shore and disappeared. His captors regret very much that he has escaped, as they were trying to ascertain his identity, and seemed greatly interested in ferreting out the matter. Nothing was heard of him yesterday.

February 4, 1883 - Chattanooga Daily Times (Chattanooga, TN)

The wild man, who is absorbing so much attention along the line of the Western & Atlantic railroad, is making toward this city. As was stated yesterday, he escaped from his captors Thursday night and was not seen again until yesterday morning, when he emerged from a forest near Tilton. Later in the day he was seen near a slaughter house feeding on a dead carcass. An attempt was made to capture him, but he fled and again disappeared. His raiment now consists of the remnants of a coat and a pair of socks. His long beard has been partially torn out by the roots and one side of his face is bare. This intensifies his hideous appearance. Dalton is excited over his approach and will give him a cold reception.

February 5, 1883 - Chattanooga Daily Times (Chattanooga, TN)

The wild man is coming this way. He was seen near Ringgold yesterday afternoon.

February 7, 1883 - Summerville Gazette (Summerville, Chattooga County, GA)

The people living along the line of the W. & A. railroad are excited about a wild man. He was first noticed near Kingston, feasting on a dead horse. He passed through that place, we believe, wearing nothing except the horse's hide over his shoulders. He was captured near Cartersville. A negro was hired to take him into the river and wash off the blood of the horse. He knocked the negro down, swam to the other side, and escaped. He was next seen near Tilton, feasting on a dead carcass. He had on then a remnant of a coat and a pair of socks.

February 8, 1883 - North Georgia Citizen (Dalton, Whitfield County, GA)

The wild man has been giving the good people of Catoosa county a fright.

February 10, 1883 - Chattanooga Daily Times (Chattanooga, TN)

TRENTON, GA., Feb. 9.—The wild man, as he is called, has just passed here. He seems to be no more than a crazy man, and I should think there is not much harm in him. He has gone on in the direction of Rising Fawn.

February 11, 1883 - Atlanta Constitution (Atlanta, Fulton County, GA)

DALTON, February 10.—Last Wednesday morning early a long caravan of cotton wagons loaded with cotton from the plantation of Colonel S. M. Carter, Murray county, arrived in our city. They created quite a ripple in trade for a day or two. The procession was a novel sight—each wagon, from two to six mules drawing them, containing from one to five bales of cotton each. The wild man of the woods, or the long-haired beast-eater of north Georgia, passed through this place a few days ago on his northern journey.

February 13, 1883 - Daily Chronicle (Knoxville, TN)

TRENTON, GA., February 10.—We had the "wild man" with us today. His approach was heralded yesterday morning by Conductor Rape, who saw him about a mile north of Morganville, standing near the railroad quietly devouring a raw rabbit. He was reported by telegraph as passing Morganville about 8am. A crowd assembled at the depot here, provided with a "lasso," to capture him. After waiting several hours for his approach, the crowd dispersed, and nothing more was heard from him until this morning, where he was reported to be about a mile north of here, eating his breakfast, consisting of a muskrat, which he cooked and ate, while using water from the railroad ditch for the purpose. His is apparently about thirty-five years of age. His beard is about eight inches long, and his hair reaches to his shoulders. He carried a large sack on his back and another on his shoulder. The latter had a hole in it, which disclosed a green hide of some kind. He said he was going to make shoes of the hide and bottom them with 2x4 scantling which he carried in his hand. He said he was from Cincinatti and going to Bridgeport. He spoke only when asked a question, and answered in as few words as possible. His voice is very weak. He stayed with us only a few minutes, having been turned from his course by some boys, who invited him to go up in town. He is certainly crazy, but appears entirely harmless. No one attempted to discover what his sack contained, as it was considered dangerous (to the sense of smell) to approach that near him.

February 15, 1883 - North Georgia Citizen (Dalton, Whitfield County, GA)

The wild man has taken up the line of the Alabama Great Southern Railway, and was last seen in the neighborhood of Rising Fawn, in Dade county.

February 18, 1883 - Chattanooga Daily Times (Chattanooga, TN)

It was rumored a few days since that the wild man had been captured in Collinsville, Ala., and put in jail on the charge of vagrancy, but upon investigation there seems to be no truth in the rumor. He passed through Collinsville, and when last heard of was terrifying the inhabitants of Gadsden. His appetite for carrion seems insatiable, and wherever any can be found he preys upon it until the last vestige disappears.

Commentary:

This sighting is by far the longest trail of reports of a single individual, but it is also quite perplexing, as it seems to blur the line between relict hominoid and human. Specific language in the reports, such as "hairy demon," "yahoo," and "monster," suggests a non-human appearance, and certain features, such as 16" feet (the equivalent of a US men's size 26 shoe), seem to corroborate this. Some characteristics, such as the long, red beard; tattered, minimal clothing; and even the ill-fitting ring, are consistent with the Neanderthaloid profile. However, the later description out of Trenton, where the wild man cooks food and speaks English, is decidedly human.

The discrepancy is easily explained by the existence of two individuals: one Neanderthaloid and one human. The first individual was a Neanderthaloid who interacted with humans only when forced. The second individual was a human who willingly spoke to the people in Trenton without the need for or threat of violence. The first individual had a long, red beard on half of his face after part of it was torn out by the roots while escaping his guard in Cartersville. The second individual had an eight-inch-long beard across his full face just ten days later. The first individual was satisfied with his horse-hide garment and only kept the coat and socks of the complete outfit given to him. The second individual wore more respectable clothes and was preparing to make himself a new pair of shoes. When comparing the descriptions in this way, it is clear that the string of newspaper articles represents two distinct individuals and that the first individual neither appeared nor behaved like a modern human.

The Railroad Menace of Atlanta

Period: March of 1884

Location: Atlanta, Fulton County, Metro Area

March 1, 1884 - Macon Telegraph (Macon, Bibb County, GA)

FROM ATLANTA

[Special Correspondence]

A report was lodged at police headquarters today that a wild man was devastating the outskirts of the city, near the E. T., V and G. Railroad. He was described as huge in size, muscular, very hairy, his hair reaching considerably below his shoulders, his beard below his waist, both jet black, his complexion like an Indians and his manner savage. He has frightened several people in the neighborhood and was reported as a dangerous character. A brace of the mounted police made a raid for the wild man this afternoon and found his den and various evidences of the recent presence of some wild, uncivilized being, but were satisfied from the indications that he had left for other regions. The suburban villages need to be on the alert.

March 12, 1884 - Summerville Gazette (Summerville, Chattooga County, GA)

A wild man is reported as infesting the suburbs of Atlanta.

Commentary:

The description of this wild man is lacking in some of the features we frequently see in Neanderthaloid reports, such as the claw-like nails, but the description is definitely that of a relict hominoid. Despite his skin tone being compared to that of a Native American, he was "very hairy." Native Americans are generally lacking in both facial and body hair relative to other ethnicities, so this "very hairy" wild man who lived like a "wild, uncivilized being" was certainly not a Native American. The exceptionally long beard confirms that he was a Neanderthaloid-type relict hominoid.

The Hairy Man of Carroll County

Period: September of 1884

Location: Carroll County, West Georgia

September 12, 1884 - Carroll Free Press (Carrollton, Carroll County, GA)

We have not heard any more of the bear recently but heard of a man supposed to be wild who was seen near J. J. Knight's farm on the river. Mr. Knight told us that on the night following the evening the hairy man was seen, Mr. Thomas Huckaba crossed over the river and spent the night on the other side and that J. W. Reese ferryman at Moore's ferry would not stay at home, but spent the night at J. D. Moore's.

Commentary:

The *Carroll Free Press* had no more information on this sighting, so we must settle for the vagueness of "hairy man" as the description. While it suggests a relict hominoid, it does not provide enough information to determine what type.

The Barefoot Screamer of Tift County

Period: April and May of 1885

Location: Ty Ty, Tift County, Southwest Georgia

April 20, 1885 - Macon Telegraph (Macon, Bibb County, GA)

Mr. John J. Williams has found a wild man in the woods near his place. One day last week this man ran through his field when Mr. Williams' little boy was at work. He went along hollering and uttering a loud shrill scream. After he ran over and knocked down his fence, he was seen by the little boy creeping along behind the trees. He was watching the boy as if waiting to get a chance to pounce upon him. The boy made his escape. The man is almost without clothes and barefooted.

April 23, 1885 - Dawson Journal (Dawson, Terrell County, GA)

A wild man, barefooted and almost nude, has been discovered near Ty-Ty, in Worth county. He utters shrill cries, and skulks through the woods and fields. Efforts are being made to capture the man.

April 25, 1885 - Weekly News and Advertiser (Albany, Dougherty County, GA)

The Savannah News has discovered, by way of Macon, the regular, annual "wild man." This time he is in Worth county. "He utters shrill cries and skulks through the woods and fields," like all of his predecessors. His is probably merely some office-seeker. "The woods are full of them."

May 26, 1885 - Macon Telegraph (Macon, Bibb County, GA)

The wild man who has been lurking around the neighborhood, was seen again a day or so ago, so I hear.

June 4, 1885 - Sumner Free Trader (Sumner, Worth County, GA)

The Ty Ty correspondent to the Macon *Telegraph* must be the wild man of Worth, as we never hear of the meandering of this mysterious person, only through the *T. &. M.* If he is not he, in his exposure to climatic influences in the search of the W. M. has caused him to be a little BROWN.

Commentary:

Again, this sighting drew the attention of several newspapers, but the amount of detail leaves much to be desired. A wild man hollered and screamed, appeared to wear very little clothing and no shoes, and lurked on the edge of the woods watching a little boy. The description is consistent with other relict hominoid reports in this book, but we need more information to discern which relict hominoid type was the culprit.

It should be noted that Ty Ty was part of Worth County in 1885, as described in the report. The eastern part of Worth County was carved out in 1905 to create Tift and Turner Counties. Ty Ty now lies within the borders of Tift County.

The Cabin Dweller of Atlanta

Period: August of 1885

Location: Atlanta Suburbs, Likely DeKalb County, Metro Area

August 20, 1885 - Savannah Morning News (Savannah, Chatham County, GA)

As a *Capitol* reporter was coming down the court house steps he saw a negro making rapid strides toward the station house. The reporter followed, and saw Capt. Crim jumping on a Washington street car. The reporter stepped on the car and asked the Captain what was up, and he replied that negroes out in the northeastern suburbs of the city were in a state of excitement at the appearance of what they called a wild man.

For several days past at various times it has been reported that something "neither brute nor human," has been seen in the neighborhood. These reports caused some uneasiness, but as only seeing was believing with most of the hearers, there was no great alarm until yesterday afternoon, when three half grown negro children came in from the woods and said the wild man was in an old deserted cabin not more than half a mile away. About twenty men and women repaired continuously to the place, and when within 200 yards of the house saw a terrible looking object rush into the woods.

"And you can find it right dere now, boss," said a negro to Capt. Crim. So the Captain and the reporter, with about 100 colored guides, went tramping through the woods. Capt. Crim was skeptical and the reporter was doubtful, so neither were prepared for the startling sensation soon to be developed.

After walking little over a half mile the negroes in front came to a sudden and exclaimed in a whisper, "Yonder's the house!"

Capt. Crim tried to get the negroes to surround the cabin and make an effort to capture whatever might be in the house, but the colored posse were too badly frightened to take any steps toward acting on the offensive. So the Captain and the reporter, armed with stout sticks, slowly approached the open doorway of the hut. When within 150 yards of the house, the reporter saw something move within, and he grabbed the Captain by the arm and said "Watch out, Captain!" Then came a horrible yell and a startling apparition dashed out of the doorway, and, turning to the right, disappeared in the woods.

The object seen certainly had the shape of a human being, but the body was covered with long, shaggy hair, and the feet and hands resembled claws. Capt. Crim and the reporter are both of the opinion that the object they saw is a wild man—a human being living in the woods like an animal.

A party has been organized who will go out for the purpose of trying to capture the strange being.

Commentary:

This account is heavy on the suspense but light on the details. Still, we can be assured this was, in fact, a relict hominoid, and there are several clues to suggest that it was a Neanderthaloid type. First and foremost, both Captain Crim and the reporter for the *Atlanta Capitol* newspaper believed the creature was human. This confusion is ubiquitous among the Neanderthaloid reports in this book. Additionally, the hands and feet are said to resemble claws, a trait also found in several Neanderthaloid reports.

The River Beast of Butts County

Period: November of 1885

Location: Jackson, Butts County, Central Georgia

November 30, 1885 - Savannah Morning News (Savannah, Chatham County, GA)

For some time past persons coming to this town from a westerly direction have reported seeing a strange animal, but just what it is no one has been able to determine. It has been variously described and discussed, until many people only travel through the section of country west of here with fear and trepidation. The unknown animal is said by some to be as large as a mule, and when pursued or disturbed utters a piercing scream like unto that of the whistle of a steam engine.

The beast readily takes to water, swims with marvelous rapidity, and at intervals, where the water is deep enough, dives beneath the waves out of sight and remains under for quite a time. The reports as to the appearance of the beast differ very much. One gentleman who claims he saw it at close range says it had a shaggy coating of hair, a large head, and ran with great speed. Another account states that while the animal sped through the water with amazing swiftness, when it came to the shoals its gait changed to a tumbling locomotion, similar to the progress of a seal.

Mr. Beauregard Heath, who lives near Flat Shoals, while walking along the Towaliga river recently, was suddenly startled to see a strange monster shooting through the water, snorting and blowing in a manner that was appalling. The water where the creature was first seen by Mr. Heath was deep, but immediately on ahead was quite a stretch of shoals, where the water was shallow. When the animal reached this portion of the river so great had its momentum grown that it shot upwards and out of the water, landing with a splash on its side. Then began its tumbling, pitching gait. Though it was a most awkward motion, its speed was far greater than an ordinary fast walk. In less time than it takes to tell it, the amphibious animal was over the shoals, and with a shrill cry, disappeared in the deeper water beyond. When some 20 feet from where it shot under the water, it rose again, with a savage puff, and shaking the water from its huge head, went careening on its way. A bend in the stream hid it from Mr. Heath's gaze. In conversation, Mr. Heath said it was as large as a good sized mule, and weighed he should judge, 600 or 700 pounds.

Other reputable gentlemen have seen it, and all are unanimous in saying they never saw anything which even resembled it. None, however, took the time to study its looks carefully, for when it presented an appearance they had urgent business in other quarters. A darkey seen it last week while out hunting. He was in the woods sneaking on a squirrel to get a shot, when, turning at the crackling of a branch, he saw behind him an animal, which he said looked like a lion, a bear, and a monkey. Having no desire to post himself up in natural history—being no naturalist, and not caring to find out to what beast family the uncanny thing actually belonged, he took a walk. As soon as he got the proper use of his limbs, he fairly flew through the woods—over stumps, through briars, across gullies, until he struck the dirt road, and was at home. A *News* reporter asked the negro about the circumstances yesterday, and further asked him to describe the animal's looks. "Fo' God, boss, I nebber stopped to 'xamin' hit closely. Jest as I wuz makin' up my mind to look at hit, hit showed hit's teeth, and I left in a hurry."

Mr. O. F. Williams, who came into town yesterday, when crossing the Towaliga river, saw a strange looking object swimming the river some distance away from him. He could not determine what it was, having never seen anything resembling it. On his way to town he met Mr. Willis Evans, accompanied by several other parties, who stated to him that they were going in quest of the unknown beast which had been terrorizing the people along the Towaliga river. Mr. Williams then recalled to mind the strange creature he had seen swimming the river, and he spoke to the party about it. Breathlessly they informed him that that was what they were in quest of, and leaving him they hurried on towards the river. The result of their hunt will be anxiously looked for, and it is to be hoped that they will be successful.

What sort of beast it is no one can even conjecture. It is possible that it is some animal which has escaped from a strolling menagerie, and though a stranger in a strange land, has managed to live and thrive. The *News* will keep its readers posted, and if the unknown creature is captured our readers will be fully apprised of the event, together with the most minute particulars pertaining thereto.

December 1, 1885 - Macon Telegraph (Macon, Bibb County, GA)

The Jackson News contains a lengthy account of a strange animal seen by many reputable gentlemen disporting on the banks of Towaliga river, and in the water in the neighborhood of Flat shoals. It is described as being about the size of an average mule, and with all the characteristics of a seal. It travels best in the water. Numbers of citizens have been frightened by it, and on Saturday a posse of men went in search of it. Some seem to think that it is a seal that has escaped from some traveling menagerie.

Commentary:

Like many reports in this book, this creature draws comparisons to animals that it simply cannot be. Mules don't swim, and seals don't run, but relict hominoids do both. Key characteristics of this hominoid are the "shaggy coating of hair," the awkward gait, the ability to swim and run at a fast pace, and teeth-showing behavior.

The shaggy hair is consistent with the Skunk Ape profile, and witness accounts of Skunk Apes suggest that, unlike the known species of pongids, they are quite capable of swimming. The quadrupedal running of the Skunk Ape and other pongids easily explains the awkward tumbling gait described by one witness. Like the seal of the account's comparison, pongids will support their weight on their arms to swing the pelvis forward. For rural Americans who had not seen a gorilla or chimpanzee, the seals of traveling circus menageries would provide a suitable analogy for this typical pongid trait.

Another typical pongid trait included in this creature's description is the flashing of teeth. This is sometimes described as a grin, but it is not a display of joy. Gorillas and chimpanzees are known to flash their teeth as a defense mechanism when they are stressed or afraid. As it turns out, the Skunk Ape was as scared of the human as the human was of the Skunk Ape.

The Sandwich Cannon of Dade County

Period: April of 1886

Location: Morganville, Dade County, Northwest Georgia

April 26, 1886 - Chattanooga Daily Times (Chattanooga, TN)

Yesterday afternoon while Arthur Thacher and Gus Krause were riding in the vicinity of Crawfish Springs on bicycles, they were startled to behold a man of extraordinary mien and remarkable appearance near the road side. They were about two miles this side of the springs and had turned a sharp bend when they suddenly came upon the strange spectacle. The man was to all appearances forty or forty-five years of age and perhaps five feet six inches in height. He was barefooted and bare-headed; his clothes were in shreds; he was coatless and almost shirtless and the shreds of his trousers were knit together with leaves and twigs; his hair fell below his shoulders and his beard came below his waist. His arms were bare and blistered by the sun, and his entire body was covered with briars and brambles. The young men came upon him unexpectedly, and were considerably startled by the strange appearance of the man and involuntarily ejaculated, "Hello!" The man quickly turned in a frenzy of fright, his eyes fairly glistening, and when he saw the bicycles he gave a frightful shriek and in a second had disappeared in the dense forests through which the road passed. The young men made diligent inquiry to ascertain something in reference to this creature, but no one living in that vicinity had ever heard of him before. It is thought he is the wild man who was seen on the side of Lookout some weeks ago.

April 27, 1886 - Chattanooga Commercial (Chattanooga, TN)

The Chattanooga Times gave an account yesterday morning of a wild man seen near Crawfish Springs by Arthur Thatcher and George Krause. A Commerical reporter was detailed to hunt it up and found it fully reliable.

The man after an exciting chase was captured near Morganville. He is covered with hair, as is an animal, and had a few remnants of clothing hang to him. He escaped from the capturing party and upon being pursued it was found that he had joined a singular looking creature walking on all fours and completely covered with hair, while a long mane dragged the ground. Both were then taken by means of a lasso by John Simpson and Harry Whitley and it was found that the other creature was a wild woman. She had been wild so long that she walked on all fours. Her nails have grown so long that they appear like claws, and she uses them for the purpose of climbing trees. Neither of these strange beings can speak but give vent to unearthly sounds resembling no animal that is known. They are supposed to be Annanias and Sapphira Wilkinson who disappeared from the western part of North Carolina some years ago. They are now securely fastened in a cavern and fed by means of a cannon loaded with sandwiches which is fired at them at regular intervals. When sufficiently tamed they will be sent to the Northern dime museums, under the management of Sam Williard.

April 28, 1886 - Chattanooga Commercial (Chattanooga, TN)

The publication in the Commercial yesterday that the Times story of the discovery of a wild man near Morganville had been corroborated created much excitement. The additional fact that a wild woman had been found heightened the excitement to such an extent that a number of people did not go to bed last night. The sympathetic portion of the community expressed great delight that a wild woman had been discovered. They said the wild man would not now be so lonesome.

The Times reporters were very much agitated over the new discovery made by the Commercial. Preparations were made at the office of the Market street organ of the mugwump Democracy all night for the departure of the flood reporter of that sheet to the cave where the wild couple are confined. The flood reporter of the Times was selected because a brave man was wanted. He had met so many dangers during the flood and had walked into the very jaws of death by falling out of his "specially chartered" boat into the water which measured three inches in depth and barely escaped drowning. He took the first train for the place where the wild people are confined, armed with three shot guns, seventeen pounds of dynamite, four packages of fire crackers, a dirk, one steel trap, a bottle of chloroform, a pair of nippers and a lead pencil. If that is an impossibility, then they are to be photographed. Should that prove a failure, a likeness is to be taken of a cannibal, the photograph of which was telegraphed from New York on Monday night.

On the return of the reporter, some startling revelations may be expected unless he should fall while ascending the mountain and break his bottle of chloroform. We shall patiently await the thrilling story.

April 30, 1886 - Chattanooga Commercial (Chattanooga, TN)

The announcement in Wednesday's Commercial that the flood reporter of the Times had gone to Morganville to search for the wild man and woman who had been discovered in the mountains has caused much anxiety among the friends of that plucky newspaper man. Up to last evening he had not been heard from. There was a feeling that he might have suffered death at the hands of the moonshiners, but being so heavily armed, it was thought he could put a whole army with banners and moonshine whisky to flight.

A special dispatch received by the Commercial this morning stated that the young man was seen, last evening, wandering around on the mountain hunting the cave where the wild couple were confined. The objects of his search will probably be found today and the thrilling story of his narrow escape and remarkable tramp will appear strictly in book form. It will be prepared expressly for sale on the trains of the different railroads.

Commentary:

The t-shirt guns of modern sporting events have nothing on the sandwich cannon of 1886. It almost outshines the wild couple themselves, which is a difficult feat as this is our only definitive sighting of a wild woman.

Though the woman seemed to walk on all fours, the man and woman were Neanderthaloid-type hominoids. The man was approximately the same height as a human, with long head hair, a long beard, and a covering of body hair that collected briars and brambles despite his smattering of ragged clothing. The woman was also covered with body hair and had long head hair. Given that this is mountainous territory and a group of humans was chasing them, the woman seemed to be walking on all fours because she was using her hands to help her navigate the steep terrain. While the Pacific Northwest Sasquatch has midfoot flexibility, which aids them in climbing steep, mountainous terrain, the Neanderthaloid has a fixed arch like that of a modern human. Thus, a Neanderthaloid may have to grasp rocks or small trees to scale rough terrain, giving the appearance of walking quadrupedally.

Beyond the relict hominoids themselves, this report is also a prime example of how ridiculously relict hominoids generally and Neanderthaloids specifically were treated. The writers and witnesses felt that this couple, supposedly Annanias and Sapphira Wilkinson, were human, who had grown hair all over their bodies and lost the ability to speak. Given our current understanding of hirsutism and hypertrichosis, this is absurd; however, it was the commonly held belief of the origin of wild men at the time. And yet, despite their belief that the couple were, in fact, human, Mr. Simpson and Mr. Whitley thought a sandwich cannon, dynamite, and chloroform were the best way to interact with the couple.

The Cow Man of Butts County

Period: June of 1888

Location: Jackson, Butts County, Central Georgia

June 5, 1888 - Middle Georgia Argus (Indian Springs, Butts County, GA)

In going over my plantation a few days ago I saw some cow tracks. The tracks were in a place, strange to say, that a cow could not possibly have gotten into. Moved by curiosity I followed the tracks, and directly I came upon a—I don't know whether you would call it a wild man or a beast. It had hoofs like a cow, hair like a cow, a short tail something like a deer, hands like a man and extra large red eyes. It walked like a man bent half forward. On getting a scent of me, it gave a peculiar whistle like a deer, raised its head and dashed away. I was greatly excited, but as soon as I recovered sufficiently I went to see my neighbor, Joe Wright. He asked me not to say anything about it, as he thought he could get a large steel trap with wings like a partridge net for $100 and catch the creature. In case this plan failed Mr. Wright said he would offer a $500 reward for it alive, or $150 for its hide. Mr. Wright already has a circus wagon, and says that if we could get this curiosity our fortune would be made. Neighbor Wright is a married man and didn't move fast enough for me, for, honestly, I am afraid to move about my premises. I went to Marshal Malone, but he didn't give me much encouragement. Then I went to Marshal Coon. He said that he was "no detective, and did not consider it his business to tackle a wild man in the woods, but if it came to town and cut up any devilment, he would carry it before the mayor." I spoke to the sheriff about it and he said he would do all in his power to capture it, and, with this object in view, has ordered a fine bloodhound. He wants 150 men to meet him in Jackson on the fourth Sunday in this month, but as for myself I am going to stay at neighbor Ball's till that animal is either caught or run out of this country. I think it will be very dangerous for black folks to gather blackberries until it is caught or killed.

—Jim Swinger

Commentary:

It's hard to say what poor Jim Swinger saw on his farm. A relict hominoid shouldn't make tracks like a cow, and a cow shouldn't walk upright like a man. Mr. Swinger was terrified, so that fear surely distorted his description; however, there is enough information for us to conclude that the cow-man was a relict hominoid. Three aspects of this story are of particular importance: the "peculiar whistle," the large, red eyes, and the stooped posture like a "man bent half forward."

Whistling calls are commonly reported among the Pacific Northwest Sasquatch type and less commonly, but still regularly, among the Eastern Bigfoot type. On the other hand, whistling calls are relatively rare in known wildlife, so there is little opportunity for misidentification.

Red eyes are also commonly reported in relict hominoids. While great apes and humans do not have the *tapetum lucidum* membrane that creates the eye shine in felines, canines, and other nocturnal animals, it is suspected that both the Pacific Northwest Sasquatch and the Eastern Bigfoot have eyes substantially larger than those of modern humans, primarily due to their size. The increased pupil size would allow more light to reflect off the blood vessels in the back of the eye; thus, their eyes seem to glow like the devil... or like every photo from the 1990s.

The stooped posture harkens back to the leaning of the Slant-Eyed Giant of Cherokee lore. This posture has likewise been reported in the Eastern Bigfoot type; thus, it would seem that Jim Swinger saw an Eastern Bigfoot. Why he thought it had hooves like a cow, though, is anyone's guess.

The Fort Dweller of Harris County

Period: June of 1888

Location: Mountain Hill, Harris County, West Georgia

June 8, 1888 - Hamilton Journal (Hamilton, Harris County, GA)

Harris county is absorbed in the sensational mystery which was unearthed near there yesterday. Half a mile west of this place, on a high and rugged peak of the Pine mountain, is an old Indian fort built of stone, and, owing to the ruggedness of the mountain, approachable on only one side. Its high, thick, stone wall, and its peculiar situation, made it quite a stronghold for the Indians in time of war. For many years it has been remembered by only a few hunters and older settlers, so deeply was it buried in the mountains. Recently, however, the neighbors who had been attributing the loss of live stock to the sheep-dog and the deft handed darky, have had cause to suppose the old fort to be inhabited by thieves.

Tuesday night Mr. B. G. Whitten passed near the mountain on his return from a neighbor's house, and heard the piteous cries of one of his goats as it seemed to be carried up the mountain side to the old fort on the summit. Investigation disclosed signs of habitation and resulted in a band of half a dozen neighbors yesterday forcing an entrance and capturing a man in the most abject state of barbarism. He showed fight and was felled to the ground twice with the butt of a gun. With great difficulty his hands and feet were securely tied. He had no weapons and was dressed in nature's garments. A thick stubby hair has grown all over his body, his full, black beard reaches to his waist and his long unkempt hair hangs down his back and about his shoulders. He has keen black eyes that seem unaccustomed to the light of day, and since capture he has not uttered a sound. He seems a man of forty years old, six feet two or three inches in height, and weighs about two hundred pounds. There is not a pound of surplus flesh about him—all is bone and muscle—and his strength is marvelous. His hands are talons and his india-rubber feet scarcely resemble the feet of a human. He was carried by the sheriff this morning to Hamilton, where it is thought Ordinary Williams will adjudge him insane, and have him sent to the asylum.

Who this mysterious person is or how long he has been cut loose from civilization no one knows. Some of the older settlers tell of a young man, Redmond Sykes, who dazzled this part of the county away back in the sixties. He is described as tall, slender and straight, with intensely black hair and eyes, silken black mustache and soft white skin. His dress was scrupulously near, and he was the picture of a handsome man. He was a man of leisure, had no visible means of support, but was never without money. He claimed to have a plantation and slaves in Virginia and one in Mississippi. For two years he was the star of aristocratic society that adorned this section, and most fathers looked on his attentions to their daughters with favor. One afternoon he called to see a lady for whom he had many times showed a preference. For two hours the interview lasted, and he left wearing a stony look of despair. His landlady noticed that he refused to attend supper as usual, and that night he disappeared and was never seen by the people of this place any more.

Those who remember the young man are inclined to think that they have found him in the person of the prisoner captured yesterday at the old fort. However, all this is merely conjecture, and it is doubtful whether or not the real truth will ever be known and the mystery cleaned up.

—Lum Duke

June 26, 1888 - The Free Lance (Fredericksburg, VA)

The usual wild man has been found in the mountains of Georgia.

Commentary:

This report is the gold standard for Neanderthaloid types in Georgia. Like the Sandwich Cannon couple, it is also an example of how Neanderthaloids were always categorized as feral humans by those who encountered them. Those who were captured were shipped off to freak shows or lunatic asylums, and the relevant scientific institutions were never informed.

Typical of the Neanderthaloids, this hominid was more robust and muscular than the standard human. He had long head hair past his shoulders, a beard that hung to his waist, and a coating of thick hair all over his body. He did not speak but lived in the shadow of humanity, occupying an abandoned stone fort and stealing livestock from local farmers. Like many other Neanderthaloids, the appearance of his nails and fingers seemed to resemble claws or talons. All are characteristics that confirm he was *not* Redmond Sykes.

The description of the eyes is quite interesting as it confirms the theory behind the glowing, red eyes discussed in the previous account. This hominoid's eyes were described as large, black, and "unaccustomed to the light of day." A highly dilated pupil easily explains these characteristics. That dilation would provide a substantial window for light to reflect off the blood vessels in the back of the eye.

The Farming Bear of Jones County

Period: July of 1889

Location: Jones County, Central Georgia

July 26, 1889 - Savannah Morning News (Savannah, Chatham County, GA)

Out in the Oaky woods the wild animals frequently exhibit intelligence to a remarkable degree. One of the largest planters, J. K. Beal, in going the rounds of his immense plantation noticed that in a field bordering the Pocasin swamp the best rails from his fence had been abstracted. Any rails decayed or defective were left, while only the best rails had been taken. This continued for quite a while. The planter supposed that the negroes had been taking them, but could find no clue to the thieves. After awhile his hogs began to disappear, until at last when he went to call them up none responded to the cry. Next, in the same field from whence the rails had been abstracted, the growing corn crop began to suffer. Bushels of roasting ears disappeared, and at last the tracks of some large animal were discovered leading from the cornfield to the mysterious depths of the Pocasin. They were the huge tracks of a bear. So much damage was being done to the crop that it was determined to try to find the brute. Accordingly a party was organized and away they went, following the trail that became more and more indistinct. The Pocasin swamp is interspersed with dry hammocks like oases in a desert. On, on, through mire and ooze, on through the little shaded islands, and back again to swampy ground. At last they reached a large, high hammock rising almost like a hill from the mysterious depths of the swamp. Here high and dry they were astonished to discover a great pen, "as large," says the narrator of this singular adventure, "as large as a great house." Upon examination it was found that the pen had been built of the stolen fence rails. Within it were the stolen shoats sleek and fat. Around it in every direction were the tracks of the bear innumerable in number and the ground well trodden down. Within the pen were the remains of some of the roasting ears abstracted from Mr. Beal's field. The hunting party were struck with amazement, and sat down around the well filled pen to try to unravel the mystery. There was but one solution. Every evidence pointed to this one fact. The bear had stolen the planter's fence rails, and built the pen upon the secluded hammock, had stolen his hogs, and then selecting his best shoats had penned them up and was fattening them up for winter use upon Mr. Beal's roasting ears. The hogs were recovered, taken to Albany and sold readily for from $7 to $8 each to a local butcher. It is said that the butcher's customers were so delighted with the sweetness of the pork and its juicy richness that they clamored for more of the same kind, but hogs fattened by bears upon roasting ears were an uncommon commodity, and the meat could not be duplicated. An Albany saloon man applied to the planter for one of the toothsome fat shoats. He wanted to barbecue it for his free lunch counter. But the supply was exhausted and the free lunch fiends were deprived of this unusual dainty. On the Fourth of July, while H. H. Savage was spending the day hunting he killed in the Pocasin a bear which turned the scales at exactly 400 1/2 pounds. It is thought that this is the bear that tried its paw so skillfully in the business of raising meat.

August 9, 1889 - Crawfordville Democrat (Crawfordville, Taliaferro County, GA)

One of my good old neighbors, "Grand Pa" says that near the Big Swamp, a man had a corn field and he kept missing his hogs until they were all gone.

Then he began to miss his corn.

He got a posse of men and made a search and to his surprise, the hogs were found in the swamp and the corn was toted to the hogs by bears.

The bears cut the timber, split the rails and built the pen. Them were bears they were.

Commentary:

As mentioned by Jeffery Wells in his book *Bigfoot in Georgia*, bears do not farm. They cannot farm. They lack the capability for strategic thought and the hand structure to make this possible. Of course, these people knew that. The *Crawfordville Democrat* report reeks of sarcasm in its insistence that the true culprits were bears.

Ironically, this account is the opposite of what we usually wrestle with in the cryptozoological community. Typically, people misidentify known animals as cryptids. Bear sightings represent a significant portion of the Bigfoot reports made annually, and we must constantly sort bear tracks out of the Bigfoot footprint evidence. To have an instance where relict hominoid behavior and footprints are mistaken for those of bears is quite uncommon. At the very least, it suggests that the tracks were unusual in appearance; otherwise, the report would have assumed a human livestock thief.

Given the bear confusion, the most likely culprit would be an Eastern Bigfoot type. A Neanderthaloid-type track would likely be indistinguishable from that of a human other than in size, as the Neanderthaloids are noted as having the same fixed arch seen in modern humans. A Skunk-Ape-type hominoid would have a track with a divergent big toe, which is quite different from a bear track with its equally sized and splayed toes. Alternatively, an Eastern Bigfoot track would be more similar to that of a Pacific Northwest Sasquatch. With the slight splay of the toes and the ability to produce a half-track, thanks to a flexible midfoot, an Eastern Bigfoot track is the best candidate in Georgia for being mistaken with a bear track.

If that is the case, it would speak volumes to these creatures' intelligence, which certainly outpaces our current conception of relict hominoids as giant, bipedal apes. Then again, we are constantly learning about the true capabilities of the great apes thanks to the spear-throwing, cave-dwelling West African subspecies of the chimpanzee.

The Tree Climber of Liberty County

Period: April of 1895

Location: Liberty County, Southeast Georgia

April 20, 1895 - Atlanta Constitution (Atlanta, Fulton County, GA)

There have been several well authenticated reports of a "wild man" in Liberty county. The negroes there assert that the said "wild man" has been frequently seen in the tops of trees at nightfall, and that his hair is as long as a woman's and almost trails the ground. Stories of whitecaps have also resulted from it.

Commentary:

Whitecaps and ku-kluxers were slang terms for members of the Ku Klux Klan. The first iteration of the Klan was founded in 1865 as a decentralized organization involved in anti-Reconstruction activity, both against black citizens (primarily those who were political activists) and the government. The notorious cross burnings and public parades did not begin until the second iteration of the Klan, which was founded in 1915. The government began to suppress the various local chapters of the early Klan in 1871, so any Klan activity occurring at the time of this sighting would have been covert. In addition to their anti-Reconstruction and anti-black activities, the early Klan was also involved in vigilante community policing. For example, several accounts exist in contemporary newspapers of Klan members assaulting men who physically abused their wives. Because of the vague nature of the report, it is unknown whether the Klan thought that the black witnesses were causing trouble, the Klan thought that the wild man was a danger to the community, or the newspaper was exaggerating and sensationalizing that aspect of the story.

As for the wild man, the story is again vague; however, the tree climbing, particularly to the tops of the trees, and the long head hair suggest this is a rare sighting of a Proto-Pygmy-type relict hominoid in Georgia.

The Woolly Monster of Wilkes County

Period: August of 1896

Location: Washington, Wilkes County, East Georgia

August 13, 1896 - Savannah Morning News (Savannah, Chatham County, GA)

Washington, Ga., Aug. 12.—The colored people living on Little river, about six or seven miles below here, are in a great state of excitement because of some large animal, which they say, is in that section of the county, devastating crops, killing and eating hogs and sheep. Several claim to have seen this terrible monster, and they say it is as big as a cow and very woolly, and that it makes a bellowing noise like a bull. All unite in saying that no matter what kind of animal it is, it has done great damage to crops.

Commentary:

This report does not explicitly describe a wild man or a being having human-like characteristics, such as walking on two legs. Regardless, the account likely represents a relict hominoid. The only possible misidentification would be of a black bear, but the size alone makes that dubious. The average cow is about five feet at the shoulder, but the average black bear in Georgia is only about three feet to the shoulder. Alternatively, an Eastern Bigfoot, larger than a black bear in height and weight, would be more comparable to a cow. Woolly is also used quite frequently to describe the relict hominoids in Ivan T. Sanderson's book *Abominable Snowmen*, and the bellowing of a bull is similar to some of the vocalizations in the famous Sierra Sounds recording of Pacific Northwest Sasquatches.

The Devil of Rockdale County

Period: November of 1896

Location: Conyers, Rockdale County, Metro Area

November 21, 1896 - Conyers Weekly (Conyers, Rockdale County, GA)

Every hound dog in this community joined in a wild chase Wednesday night. Some dog started a trail at West View Cemetery and others joined in until twenty or more were in a mad race after some unknown animal for an hour and a half when the whole thing ended abruptly and mysteriously. The trail circled the the town and passed through street after street but the dogs never gained a sight race and quit the trail near where it started. A number of boys joined in the chase and some claim to have seen the strange animal as it glided swiftly and noiselessly over the ground well in advance of the hounds.

November 26, 1896 - Rockdale Banner (Conyers, Rockdale County, GA)

Last Wednesday night a party of possum hunters were put to their wits by the appearance of some strange animal. Some of the dogs struck a trail in the woods north of Eastview cemetery and it was not long before fifteen or twenty were in a mad chase after the animal. It is said that it passed through Center street, and after an hour's trail the dogs quit the race near where they had started.

On Saturday night it was chased again, starting at the same place as before, and after a warm race the dogs gave it up again. Several of the party saw the animal and many shots were fired at it, but all missed the mark.

Several efforts have been made this week to capture the mysterious animal, but all were fruitless.

The reports first brought to town by the hunters were treated as a mere joke and nobody thought much of it; but some of our most reliable citizens say it is true that some wild animal is in our midst, and that our sporting boys have a splendid opportunity to have some real fun.

We understand that Mr. J. E. Maddox, who lives just beyond the woods where the animal is supposed to den, says that something of the kind was in that vicinity some three years ago, but he fails to give it any name.

It is said to resemble a man in disguise and when standing up is about the heighth of a man, but when running goes on its four legs. Its color is black and its appearance indicates great muscular power, both in jaws and limbs. The boney ridges in the skull above the eyes are extremely prominent and the teeth are very large.

The above pen sketch is said to resemble the animal very much, though it does not have long hair. Up to this time the following gentlemen have had the opportunity of seeing it, and those who doubt the authenticity of the above statement may ask any of them about it: W. N. Everitt, Seab Glenn, W. B. Smith, Marian Plunket, Luther Hollingsworth, Carl Downs, Clifford Sigman, E. S. Everitt, John R. Maddox, G. W. Alexander, and others.

Commentary:

This incident is amusing as it is our one account that comes with an artist's rendering, and the creature in the rendering looks like the mascot for a knock-off Halloween-themed cereal.

When looking at the description given by the witnesses, though, several features indicate the mystery animal was likely a relict hominoid. It was similar in shape and height to a human but with a black covering and a propensity for moving about on all fours. It had a prominent brow ridge, a strong jaw, and large teeth. All of this indicates it was likely a Skunk-Ape-type relict hominoid.

WHAT IS IT?

Our Artist, After Seeing it, Gives us A Pen Sketch of
The Mysterious Animal That Has Been Prowling
Around in The Vicinity of Eastview Ceme-
tery For The Past Week.

The Berry Thief of Chattooga County

Period: June of 1897

Location: Summerville, Chattooga County, Northwest Georgia

June 30, 1897 - Summerville News (Summerville, Chattooga County, GA)

Another sensational story last week related to a terrible wild man who chased some small berry pickers home through the twilight shades. Judging from the descriptions of him he is the identical wild man who has terrified people in every part of the county though the last decade. He has the same hairy covering and prowling habits. The story of him spread like wildfire and since then it has been almost impossible to obtain any berries as the negroes are afraid to hunt them any more. The housekeepers in town have also had their suppers prepared before dark by cooks who were "'fraid to go home atter dark 'count er dat wile man."

Commentary:

No other reports of this wild man in the area seem to have survived; however, Chattooga County is the southern neighbor of the infamous Walker County. We can presume the wild man was an Eastern Bigfoot type, as this type most frequently struck fear in the populace of Walker County. Though Neanderthaloid-type relict hominoids were common in Walker County, they were presumed to be human and did not elicit such a strong reaction from the witnesses.

<h1 style="text-align:center">The Linguistic Mystery of Heard County</h1>

Period: July of 1917

Location: Heard County, West Georgia

July 9, 1917 - Americus Times Recorder (Americus, Sumter County, GA)

A strange wild, hairy creature, bearing some semblance of a man and some semblance of a wild animal, who was captured on the banks of the Chattahoochee river by Sheriff Taylor of Heard county, has been brought to Atlanta and turned over to the federal authorities, but they have no use for him and don't want him and it looks as if Sheriff Taylor will have to take him back and feed him at the expense of Heard county.

So far as Atlanta's leading linguists can discover, he speaks no language. Dr. F. E. May, the French consul, who is able to converse in most of the civilized tongues, has tried him out with no success. "He understands nothing I speak, he speaks nothing I understand," says Dr. May.

The stranger is little and bent and carries a cane. His head and face are matted with hair. He looks like a wild man, but he appears to be perfectly harmless.

"The government doesn't want him, I don't want him, and he won't run rabbits, so what am I going to do with him?" asks Sheriff Taylor.

July 11, 1917 - Americus Times Recorder (Americus, Sumter County, GA)

The "strange wild man" captured on the banks of the Chattahoochee a day or two ago may be well known in his own bailiwick, but inadvertently strayed off his reservation. That often happens right in this section.

Commentary:

This account may represent one of the last Neanderthaloid types in the state. While described as hairy and resembling a wild man, he is presumed by the reporter and the authorities to be human. This, along with references to his facial hair, would indicate that the Neanderthaloid type is most likely if he is a relict hominoid.

The most compelling clue that this wild man was a relict hominoid is that he underwent linguistic testing by the French consul, a highly educated polyglot. This is the only potential relict hominoid in the state who underwent the same level of examination, and the result of the assessment is that the wild man spoke no known language. As such, it is implausible that the wild man was an American of European descent. There is a marginal possibility that the wild man was a Native American, the last of a remnant population that had survived almost a century past the Indian Removal era; however, this is such a slim chance that it seems less likely than the existence of relict hominoids in the area. The removal of both the Cherokee and Creek tribes was quite extensive in this region of the state. Additionally, Native Americans are unlikely to be able to grow enough facial hair to become matted and wild-looking. Instead, it seems most likely that this sighting represents an elderly or disabled Neanderthaloid who no longer had the physical prowess to fight off his captors.

The Lunatic of Wilkinson County

Period: April of 1921

Location: Irwinton, Wilkinson County, Central Georgia

April 13, 1921 - Macon Telegraph (Macon, Bibb County, GA)

IRWINTON, Ga., April 12.—A man who is either a lunatic or an escaped convict has been reported to have been seen several times in the swamps in the southern section of the county. The first to report him was some negro boys who were hunting and their dogs got on his trail and bayed him. The boys came running up to see what it was and found what they described as the most awful looking man they ever saw. They were so frightened at his appearance that they turned and fled.

They said that he was so dirty, ragged and greasy that they did not know whether his was a white man or a negro.

A few days later a white boy was working in his father's field and saw a strange looking being approaching. Upon seeing the boy, he turned and ran towards a grave yard nearby. The boy was also frightened so badly that he took to his heels and never stopped until he found his father. His father investigated and saw the tracks of the man going to the graveyard and traced him into the swamp.

None who have seen him were able to give a good description of him. All reports agree that it was impossible to tell whether he was a white man or negro and in each case he would attempt to keep his face hidden.

April 21, 1921 - Lyons Progress (Lyons, Toombs County, GA)

Wilkinson county reports a wild man running loose. It shouldn't worry, for it has nothing on the other counties of the state. There are lots of wild men running loose everywhere.

Commentary:

This report is not obviously and definitively of a relict hominoid; however, some aspects hint in that direction. The wild man is described as looking "awful" and "strange," which suggests an uncanny valley type of experience where the wild man was close enough to human to be categorized as such but different enough to create discomfort in the witnesses. The inability to determine the ethnicity of the wild man is similar to the experience of the witnesses of the Railroad Menace, who described a skin tone similar to that of a Native American in a being far too hairy to be a Native American. These hints leave open the possibility of a Neanderthaloid-type relict hominoid in Wilkinson County.

The "Boy-Fish" of Floyd County

Period: June of 1923

Location: Pinson, Floyd County, Northwest Georgia

June 5, 1923 - Atlanta Tri-Weekly Journal (Atlanta, Fulton County, GA)

ROME, Ga., June 2.—More than a score of men searched in Woodward creek in the Pinson community of Floyd County Friday for a mysterious animal that has been seen there, while Mrs. Bud Davis is confined to her bed as the result of her encounter with the freak.

The strange water animal, discovered in Woodward creek last Friday, was seen again yesterday by Mrs. Davis immediately in front of the railroad station at Pinson "standing on a log." When Mrs. Davis screamed the animal plunged into the water and the woman fainted. She was carried to her home, where she was confined to her bed the remainder of the day. She described it as having a face like a boy, but its body covered with short, dark brown hair.

June 7, 1923 - Summerville News (Summerville, Chattooga County, GA)

While about twenty men of the Pinson community are searching Woodward Creek for the "freak" that has been reported as having been seen in the water, flopping like a boy at play, but with a body covered with fur, Mrs. Bud Davis is confined to her bed as the result of her encounter with the animal this morning.

The seal, water animal, or whatever it is, was seen by Mrs. Davis this morning just in front of the station at Pinson, "standing on a log." She screamed and ran, and the boy-fish disappeared.

Immediately a posse was formed and this time the men are determined to solve the mystery if "it" sticks its head about the water. Mrs. David fainted on the way home and had to be taken home in a buggy. She was then forced to go to bed.

She described the "thing" as having a boy's face, but with a body covered with fur, and said that it slipped from the log she saw it standing on just as soon as she screamed.

The posse is armed with everything except fishing poles and bait. They have rifles, shotguns, rocks and big sticks, and it will be a sad day for the "mystery of Pinson" when they catch up with it.—Rome News.

June 8, 1923 - Danielsville Monitor (Danielsville, Madison County, GA)

Rome.—Residents of Pinson's district of Floyd county are aroused over a story vouched for by three different persons that what appears to be a sea lion or a seal is in Woodward creek, a good-sized stream that runs into the Etowah river. The animal was first seen by a young woman, and later at different times by two farmers. One of the latter states that he managed to approach within less than 30 yards of the animal as it lay upon a partly sunken log.

Commentary:

Why anyone would label this creature a sea lion or seal is beyond me, other than to rationalize the irrational. Neither sea lions nor seals would find a freshwater creek in Northwest Georgia a suitable habitat, nor would either be readily described as a boy covered with fur. This was almost certainly a relict hominoid sighting. The comparisons to a seal, however, suggest the hair was short not just on the body but also on the head. Thus, this relict hominoid was most likely a juvenile Eastern Bigfoot type.

Chapter Seven: Misidentifications and Hoaxes

A Human in Walker County

Period: January of 1885

Location: Walker County, Northwest Georgia

January 3, 1885 - Chattanooga Daily Times (Chattanooga, TN)

The citizens living in the vicinity of High Point, in Walker county, Ga., are in a state of great excitement over the appearance of a wild man, who has been seen prowling about for several days. He was first discovered about one mile beyond Dr. Park's residence, but has since taken up his abode in the ravines and cliffs near High Point, on Lookout mountain. The man is very uncouth in appearance, his face being covered with a heavy growth of short, black beard, and his tangled, unkempt hair hangs in wild profusion about his neck and shoulders. He wears a tattered overcoat and pants, goes barefooted and hatless. The sight of any human seems to inspire him with terror and fright, and he bounds away into the brush and thickets, uttering the most piercing shrieks and cries as if in distress. A Mr. Long, living in Chattanooga valley, stated to a Times reporter that he has seen the demented man twice. An effort will be made to capture him.

January 10, 1885 - Dalton Argus (Dalton, Whitfield County, GA)

A wild man is terrorizing the good citizens of High Point, in the neighboring county of Walker.

January 13, 1885 - Summerville Gazette (Summerville, Chattooga County, GA)

A wild man is reported near High Point, Walker county.

January 23, 1885 - Macon Telegraph (Macon, Bibb County, GA)

The wild man, an allusion to whom appeared in the Telegraph recently, has again made his appearance near High Point, in Walker County. The man has lost nearly all of his tattered clothing and is almost nude. He has deserted the cave which he was inhabiting when first seen two weeks ago, and his present abode is not known. Those who have been able to get a good look at the man say he has become very much emaciated and shows the effects of exposure and lack of food. All efforts to capture him have failed. His distressful, plaintive cry is heard throughout that section almost every night.

January 29, 1885 - Walker County Messenger (LaFayette, Walker County, GA)

The wild man came out yesterday to take a peep at the sun, then returned to his hole.

Commentary:

This wild man sighting stands in contrast to the Neanderthaloid sightings included in this book and is, therefore, an excellent example of the process of elimination that must be done on any witness account. Whereas the Neanderthaloids are solid and muscular, this wild man was emaciated and struggling to survive in the woods. Whereas the Neanderthaloids have long beards that fall to their waist if not past it, this wild man had only a short beard. Although Walker County was, and perhaps still is, home to relict hominoids, this wild man was decidedly human.

The Hoax of Dade County

Period: November of 1888

Location: Sitton Gulf, Dade County, Northwest Georgia

November 2, 1888 - Dade County News (Trenton, Dade County, GA)

Last Tuesday while Mr. Marion Tatum and Fate Quinton were out sheep hunting in the "Sitton Gulf" their attention was attracted by a series of unearthly sounds proceeding from a rocky ledge some hundred yards from where they had stopped in amazement and fright, Mr. Quinton being the most adventurous of the two, proposed an investigation, which was reluctantly agreed to by Mr. Tatum. After advancing cautiously to within twenty paces of the heart rending screams they beheld a sight which chilled their blood and caused the hair of their heads to "stand on end;" never did man gaze upon a more harrowing sight than that they now faced, sitting on a projecting stone near the verge of the cliff sat a man—if such it might be called, with the skin of a hog drawn loosely around his lower limbs, savage fashion. While the body was covered with a coat of coarse sandy hair about an inch long, the beard of his face was shaggy and long, and the hair of his head was long and unkempt and matted, his long sharp teeth gnawing viciously upon a piece of putrid meat of some sort—this was the spectacle which two brave men were compelled to face, but only for a few moments since the being before them had seemed to scent their presence and was away in a few slight bounds among the rocks and disappeared from their sight after two or three piercing cries left his beholders almost petrified in their position. No one knows from whence the wild man comes and this is the first time he has been seen in this neighborhood, although a report of his presence was received from Walker county some two weeks ago, yet our citizens could not give credence to such an unnatural thing, but when two of the most respectable and conscientious men of our county are witnesses we would almost believe in ghosts should they claim to have seen one. We are to believe beyond a doubt of the existence of a wild man in community, whether the world accepts our story or not.

November 8, 1888 - Walker County Messenger (LaFayette, Walker County, GA)

I ask permission and space in you valuable paper, to reply to an article that appeared in the Dade County News in the last issue of said paper. The substance of said article was as follows: C.M. Tatum and LaFayette Quinton, while in the Sitton Gulf, near Trenton, Ga., saw a wild man who was making his meals on tainted and unpalatable meats, was hairy, and in every respect as the wild beasts that roam the forest, and who fled at our approach and would not exhibit any degree of friendship towards us, but left us in utter amazement and wonder.

We desire to say that said article above referred to was and is entirely without a shadow of proof to support it, and was maliciously gotten up by the Editor, his Devil or the Devil who furnished it for the Editor, and we say there is no truth in said article. In fact, we believe that the Editor of the News or his Devil or the Devil who furnished said article, is the wildest man we ever saw. The same originated it in his own brain when wide awake. We can not give it the credit of a dream.

—C.M. Tatum.

—J.L. Quinton.

November 9, 1888 - Dade County News (Trenton, Dade County, GA)

Our neighbor, the "Messenger" comes out this week with a bristling denial of the wild man story purporting to be from Messrs. C. M. Tatum and J. L. Quinton, but which sounds very much like a lawyer had been hired to write their reply or else took pleasure in charging us with malice in getting up the little inoffensive "fib".

November 9, 1888 - Dade County News (Trenton, Dade County, GA)

Newspaper men are often mistaken and are led to publish things not altogether true. Last week we published an account of a Wild Man supposed to have been seen by Messrs. Tatum and Quinton, but it now turns out that there was no Wild Man at all, only the gentlemen above named happened to meet off in the Sitton Gulf and each thought the other to be a wild man, and both fled in opposite directions. It was not till they had gotten together and began relating their adventures that they discovered the mistake into which they had led an innocent editor. We are always willing to correct errors, and are glad to say we were mistaken in regard to the Wild Man article of last week.

November 23, 1888 - Dade County News (Trenton, Dade County, GA)

A correspondent of the Messenger "from Eagle Cliff" believes in our wild man story since such a man was actually seen in Walker. We thought we were correct after all.

December 21, 1888 - Dade County News (Trenton, Dade County, GA)

Mr. Marion Tatum gives another account. He says, after having made a close survey of the ground, and investigating the matter thoroughly, informs us that Mr. Morrison was undoubtedly attacked by the wild man.

We thought we could convince Mr. Tatum that he had seen a wild man.

Commentary:

This account appears to have been a work of fiction. There is certainly a denial by the men supposedly involved, and there is perhaps an admission from the newspaper itself that it was a fabrication, though the tone is unclear. There is, however, the possibility that it was a real story transmitted, after which the witnesses felt they had to deny the story to avoid being accused of folly or insanity. Though this is a pretty early report in the Northwest Georgia area, local newspapers had already extensively covered sightings from states such as New York and Arkansas, so the reporters would have had plenty of material from which to pull. Because of the highly poetic language in the initial report, it certainly has the smell of a hoax. As with all of these accounts, use your discretion and discernment.

The Squalling Animal of Walker County

Period: September of 1891

Location: Walker County, Northwest Georgia

September 1, 1891 - Griffin Daily News and Sun (Griffin, Spalding County, GA)

LAFAYETTE, Ga., Aug. 31.—The strange, squalling animal of Duck creek is still in the circle. Mr. J. Williams heard it on the ridge near Dr. Underwood's upper place last Wednesday night about 12 o'clock. He first thought it to be some one calling. When he got to the door it was on the ridge. He says it squalled five or six times like a woman. The people are greatly excited over it, many of the negroes declaring that the spot is haunted by evil spirits.

Commentary:

This encounter could represent a relict hominoid, but it could just as easily be a big cat. Many will associate the sound of a crying baby or screaming woman in the woods with relict hominoids, but that ignores the more likely scenario of a big cat encounter.

For example, in his book *Bigfoot in Georgia*, Jeffery Wells references the "Comference Booger," a supposedly historic report given to the Gulf Coast Bigfoot Research Organization. The Comference Booger has not been included herein because no contemporary reports of the activity could be found and because, like this report from Walker County, it could be explained by big cat behavior. According to the GCBRO Report Database, sometime in the late 19th century, the Comference Booger hid under the porch of the Comference sisters, forcefully pushing up on the floorboards and screaming "like a woman." This was before the extermination of the North American cougar from the Eastern United States in 1900. Cougars can tip the scales at over 100 pounds and stand taller than two feet at the shoulder. Additionally, like other big cats, they pounce on their prey. Any cougar hunting small game under the sisters' porch could easily knock around a few boards and make sounds resembling a screaming woman.

The problem with immediately associating crying baby or screaming woman sounds with relict hominoids is that big cats make these same vocalizations, and both bobcats and cougars were prolific in Georgia in the 19th century.

The Baby-Footed Beast of Madison County

Period: July of 1896

Location: Five Forks, Madison County, East Georgia

July 6, 1896 - Atlanta Constitution (Atlanta, Fulton County, GA)

Comer, Ga., July 5.—(Special.)—A report reached here from Five Forks that a very frightful and ferocious beast was seen by some children in a field near that place last Tuesday evening, and that their dog has not been seen since a combat with it. Late in the evening it was seen near Harry Foot's house. Will Gholston, a colored man living at Five Forks, claims that there was something very unusual in a swamp near Harry Foot's house as he passed there about dark Tuesday night. Its track is very much like that of a baby foot, except there is an imprint of sharp toes.

In a path near that place has been found the blood, horns and a few bones of a goat. As there are no goats near that place it is supposed that this animal brought its food from a pasture near Mr. Jim Faulkner's, where it was seen the morning before, and where several pigs have been destroyed.

A crowd of men were on the hunt with guns, lights and dogs Tuesday night late, but the dogs would not do more than track to the edge of a big swamp wherein it is supposed this animal at the time was stopping.

A cow which has lately been running in this swamp came home not many days ago with the flesh on her shoulders, one of her ears and flanks badly torn, having lost a great deal of blood.

Commentary:

So often in cryptozoology, researchers must sort out the misidentification of known animals as cryptids. With relict hominoids, we must primarily differentiate between unknown hominoids and bears. This report ascribes a human-like appearance to what is most likely the hind paw print of an American black bear, as hominoids do not have the claws necessary to create what is described.

While bears have rounded fore paw prints, the hind paw print is elongated, giving a more human-like appearance. In the American black bear, the hind track is 5.5 to 9 inches long, approximately equivalent to a US children's size 6.5 to a US women's size 6 shoe. Consistent with Bergmann's rule of size according to latitude, the Georgia population of black bears would be on the smaller end of that spectrum. The small, elongated track would appear child-like or baby-like to the untrained eye, but it is entirely consistent with a known species.

Another Human in Walker County

Period: Spring of 1904

Location: Walker County, Northwest Georgia

February 5, 1904 - Walker County Messenger (LaFayette, Walker County, GA)

There is a general excitement out in the ridges over the wild man that has been sleeping in Russ Childress' barn for some time. They have been trying to catch him but have so far been unsuccessful.

February 12, 1904 - Walker County Messenger (LaFayette, Walker County, GA)

There is still a great deal of comment out in the ridges about the wild man who has been slipping around for some time. Russ Childress was in town Tuesday morning. He states that it is a positive fact. He says that the wild man slept in Ely Myer's barn on Saturday night. Mr. Myers, being an old-time farmer, had quite a number of pumpkins stored away in his barn loft which were covered with hay and fodder. He had not examined them in some time supposing they were all right. On Saturday morning last Mr. Childress was down there and they together were discussing the matter as to what move they should make to catch the man, the latter saying he had not been in his barn for some time and was inclined to think that likely he had left since the hot chase they gave him during the snow. Mr. Myers said he did not think he had been around his place, but would look around and see. So on going up on the barn loft and examining the pile of seed pumpkins, found that the seed had been taken from every single one of them. They also found the wild man's bed and the husks where he had eaten the pumpkin seed.

March 4, 1904 - Walker County Messenger (LaFayette, Walker County, GA)

The wild man who has created so much excitement in our neighborhood was captured Monday morning and is now on exhibition in Chattanooga, corner 8th and Broad Streets. He will be kept over until after the May Festival and if he fails to be identified will be sent to the zoological garden at Cincinnati.

—Waverly

March 25, 1904 - Walker County Messenger (LaFayette, Walker County, GA)

Bro. Waverly, it seems as though the wild man who was on exhibition at the corner of 8th and Market has been identified and released, as he has been seen near the coke ovens.

—Jack Slasher

Commentary:

Despite being interwoven in time with the 1904 Encounter with an Eastern Bigfoot, this account likely describes a fully human individual. Firstly, there is no description of either appearance or behavior, suggesting nothing remarkable about it. Secondly, the individual was seemingly "identified and released." Even the Neanderthaloids who were mistaken for humans were taken to jails or asylums, not released on their own recognizance.

Chapter Eight: Appendix

A Word on Hybridization and Interbreeding

While I found no direct evidence of interbreeding between modern humans and relict hominoids, I saw no evidence that precluded the notion. In the case of Neanderthaloid-type relict hominoids, I found an attitude toward them that made interbreeding and hybridization a distinct possibility.

Successful interbreeding resulting in hybrid offspring requires both the carnal act and genetic compatibility. In the case of pongids, like the Skunk Ape, the chromosomal count of 48 is incompatible with the human chromosomal count of 46. With their more human-like characteristics, Eastern-Bigfoot-type hominoids may have the necessary 46 chromosomes. Still, they were very clearly seen as non-human by witnesses, making the carnal act unlikely to occur. The same could be said of the Proto-Pygmy.

Neanderthaloids, on the other hand, are almost assuredly genetically compatible with modern humans and were considered human by all who encountered them. At that point, interbreeding becomes not a moral or religious question of humans with non-humans but a social question of humans with uncivilized humans. Would a human woman be willing to mate with a man that she believed was human even if he was excessively hairy and did not speak? Probably. After all, convicted serial killers get marriage proposals all the time. Where, or if, these hybrid children would fit into society, we can only speculate. That said, if we accept that Neanderthaloids existed as recently as 150 years ago, as the evidence suggests, we must acknowledge the possibility of hybrid offspring within that time frame.

A Brief Timeline of Georgia History

Before Colonization - There were numerous Native tribes within Georgia before colonization. Two of the most well-known were the Cherokee and Muskogee Creek, but those tribes represent just a tiny portion of the indigenous history of the state, which persisted for millennia before the arrival of Europeans. Remnants of the state's prehistoric cultures include the Rock Eagle Effigy Mound in Putnam County, Kolomoki Mounds in Early County, Etowah Mounds in Bartow County, Ocmulgee Mounds in Bibb County, and a reconstruction of the Nacoochee Mound in White County.

1540 - Hernando de Soto led an expedition that traveled throughout the South, including what is now Georgia. Several Spanish explorers and settlements followed.

1629 - The English began their campaign to control what is now Georgia. Conflicts between the Spanish and the English, using the Native Americans as allies—or pawns, depending upon your perspective—continued for almost a century.

1732 - James Oglethorpe founded Georgia, the last of the thirteen colonies, hoping to create a debtor's colony to relieve the overcrowded London prisons. Among the peculiarities of the Georgia project were early prohibitions on slavery, rum, lawyers, and Catholics, all of which were later repealed.

1733 - Georgia's flagship port, Savannah, was settled. Savannah became Georgia's first de facto capital city.

1742 - Rum was legalized.

1751 - Slavery was legalized.

1776 - Three representatives from Georgia signed the Declaration of Independence, all of which had counties named in their honor: Button Gwinnett, Lyman Hall, and George Walton.

1777 - The first state constitution was adopted. Savannah became the state's first official capital city.

1779 - The Siege of Savannah occurred. The British had occupied Savannah for a year, and American and French troops attempted unsuccessfully to regain control. It was the most extensive battle of the American Revolution in Georgia and the second deadliest battle in the Revolution overall.

1780 - The state assembly moved up the Savannah River to Augusta to be more protected. Augusta was recaptured by the British not long after, so the assembly was moved temporarily to Heard's Fort in Wilkes County. The assembly moved back to Augusta in 1781 and to Savannah in 1782.

1785 - The University of Georgia was founded.

1786 - After some back and forth, Augusta became Georgia's second official capital city.

1787 - Two representatives from Georgia signed the Constitution: William Few Jr. and Abraham Baldwin. William Houston and William Pierce were also there as representatives of Georgia but were not signers. Of the four, only William Few Jr. has no county named in his honor.

1788 - Georgia ratified the United States Constitution, becoming the fourth state to enter the Union.

1793 - Eli Whitney invented the cotton gin, quickly making the cotton industry a staple in the Georgia economy.

1796 - Louisville, a planned city named after the King of France in appreciation of French assistance during the war, was completed and became Georgia's third capital city. As the seat of Jefferson County, Louisville was slightly to the west of Augusta and, therefore, more accessible to representatives from across the state. Unfortunately, it quickly became a hotbed of malaria.

1807 - The Georgia Assembly moved to its fourth capital city, Milledgeville.

1830 - The Indian Removal Act was signed into law by President Andrew Jackson.

The 1830s - North Georgia, particularly Dahlonega, experienced a massive gold rush that escalated tensions between Native Americans and Americans of European descent.

1836 - Wesleyan College, the first degree-granting women's college in the world, was established in Macon.

1838 - The Trail of Tears expulsion of the Cherokees began. White Georgians gained access to the agricultural potential and gold of the Cherokee territory of Northeast Georgia. Between 4,000 and 5,000 Cherokee died on their exodus to Oklahoma.

The 1850s - Georgia became home to more railroads than any other southern state, which set the scene for Georgia's important role in the Confederate war effort.

1861 - Georgia became the fifth state to secede from the Union. Alexander H. Stephens, Robert Toombs, and brothers Howell and Thomas R. R. Cobb became prominent in the Confederate government. Once again, all have counties named in their honor.

1864 - William Tecumseh Sherman marched through Georgia. His army crossed into Northwest Georgia after a successful campaign at Chattanooga and continued to Atlanta and Savannah in his famously destructive March to the Sea. Sherman and his men destroyed railroads and farms, burned cities to the ground, and were responsible for over one billion dollars in damage in today's money.

1867 - The Reconstruction era under occupying leadership began. Georgia, Alabama, and Florida were part of the Third Military District, operated by the United States War Department under General John Pope. Atlanta, which had been making overtures to become the new capital for a couple of decades, was selected as the headquarters for the provisional government, followed by formal acceptance as the new capital city in the new state constitution the following year.

1870 - Georgia was readmitted to the Union.

1877 - A new Georgia constitution was ratified.

1885 - The Georgia School of Technology, now known as the Georgia Institute of Technology or, more colloquially, Georgia Tech, was founded as part of Reconstruction efforts to industrialize the South.

1892 - The Coca-Cola Company was incorporated.

1912 - High schools were added to the state school system.

1915 - The second iteration of the Ku Klux Klan was organized at Stone Mountain.

1915 - The boll weevil was introduced to Georgia. Within eight years, the cotton crop in Georgia was reduced by 50%. Rural towns were decimated, some becoming ghost towns as tenant farmers were forced to migrate to northern cities for better opportunities. The subsequent transition of farmland from cotton to cash crops left the soil depleted and susceptible to erosion. Soil that had already been damaged severely by the farming practices of the 19th century, as evidenced by Providence Canyon in Lumpkin, GA, was further destroyed.

1925 - President Franklin D. Roosevelt began to frequent Warm Springs for experimental treatments of his polio condition.

1929 - The Great Depression began.

Map of Relevant Counties

1. Bartow County
2. Butts County
3. Carroll County
4. Catoosa County (Pronounced kuh-TOO-sa)
5. Charlton County
6. Chatham County (Pronounced CHAT-uhm)
7. Chattooga County (Pronounced chuh-TOO-guh)

8. Dade County
9. DeKalb County (Pronounced duh-KAB)
10. Fannin County
11. Floyd County
12. Fulton County
13. Greene County
14. Habersham County
15. Harris County
16. Heard County
17. Jones County
18. Liberty County
19. Madison County
20. Polk County
21. Rabun County (Pronounced RAY-bun)
22. Rockdale County
23. Tift County
24. Walker County
25. Whitfield County
26. Wilkes County

Recommended Resources

- Arment, Chad - The Historical Bigfoot

- Coleman, Loren - Bigfoot!: The True Story of Apes in America

- Coleman, Loren and Patrick Huyghe - The Field Guide to Bigfoot, Yeti and Other Mystery Primates

- Haywood, John - The Natural and Aboriginal History of Tennessee

- Kilpatrick, Jack F. and Anna G. - Friends of Thunder: Folktales of the Oklahoma Cherokees

- Knight, Lucian Lamar - Georgia's Landmarks, Memorials, and Legends

- McQueen, Alexander Stephens - History of the Okefenokee Swamp

- Mooney, James - Myths of the Cherokee

- Mooney, James - The Sacred Formulas of the Cherokees

- Napier, John R. - Bigfoot: The Yeti and Sasquatch in Myth and Reality

- Petolicchio, Louis R. - The Wild Man of North America: Historic Newspaper Accounts About Encounters with Wild Men, Feral Humans and Other Curiosities

- Pruitt, Matt - The Phenomenal Sasquatch: Seeking the Natural Origins of a Cultural Icon

• Sanderson, Ivan T. - Abominable Snowmen: Legend Come to Life: The Story of Sub-Humans on Five Continents From The Early Ice Age Until Today

• Souliere, Michelle Y. - Bigfoot in Maine

• Steed, Hal - Georgia: Unfinished State

• Weatherly, David - Peach State Monsters: Cryptids & Legends of Georgia

• Wells, Jeffery - Bigfoot in Georgia

About the Author

Erin Cain is a preternaturalist, exploring the strange and unexplained through her work on the What in the Sam Hill?! Podcast and her new book Georgia Wildmen: Exploring the History of Relict Hominoids in the Peach State. With a decade of experience as an electrical engineer, she brings an analytical mindset to her life-long passion for anomalous phenomena. Erin enjoys camping, collecting old books, and singing while she cooks. She lives with her husband and children in rural Georgia.

Read more at https://www.erin-cain.com.

www.ingramcontent.com/pod-product-compliance
Lightning Source LLC
Chambersburg PA
CBHW060924140726
47996CB00001B/366